D1515220

Grandma Says

Weather Lore From Meteorologist
Cindy Day

NIMBUS
PUBLISHING

Copyright © 2012, 2013 Cindy Day

All rights reserved. No part of this book may be reproduced, stored in a retrieval system or transmitted in any form or by any means without the prior written permission from the publisher, or, in the case of photocopying or other reprographic copying, permission from Access Copyright, 1 Yonge Street, Suite 1900, Toronto, Ontario M5E 1E5.

Nimbus Publishing Limited
3731 Mackintosh St, Halifax, NS B3K 5A5
(902) 455-4286 nimbus.ca

Printed and bound in Canada
NB1111

Author photo: Alex MacAulay
Illustrations: Melissa Townsend
Cover design: Heather Bryan/Jenn Embree

Library and Archives Canada Cataloguing in Publication

Day, Cindy, 1964-, author
Grandma says : weather lore from meteorologist Cindy Day.

Reprint. Originally published: Halifax, NS : Nimbus Pub., c2012.
ISBN 978-1-77108-085-9 (pbk.)

1. Weather—Folklore. 2. Weather forecasting—Popular works. I. Title.
QC998.D39 2013   551.63'1   C2013-904738-7

Nimbus Publishing acknowledges the financial support for its publishing activities from the Government of Canada through the Canada Book Fund (CBF) and the Canada Council for the Arts, and from the Province of Nova Scotia through the Department of Communities, Culture and Heritage.

Preface

· · · · · ·

D o you ever wonder what the birds would think if they realized how closely we watch them? Or what the cows would do if they knew we planned our day by whether they were standing or lying down? It makes you wonder if anyone or anything is watching our behaviour. Watching, observing, learning— that's what our ancestors did.

Today, so many of us walk around without actually looking. That wasn't the case with Grandma. She didn't have a Blackberry or an iPhone, she wasn't texting or tweeting, she was admiring and learning from nature. I think it's safe to say that the key to Grandma's prolific prognostication was her constant observation. That's really how weather lore came about: generations of people observing nature and the changes taking place all around them, then sharing that information with others. It's that "sharing" part that inspired me to write this book.

Surely we can't allow centuries of careful observation to be lost. I can't imagine a generation growing up without the whimsy of weather lore. I often meet people who remember hearing their parents or grandparents talk about the "sugar snow" or the "mackerel sky" but they can't quite remember what it means. For you, I hope this collection of weather lore will bring back pleasant memories of a simpler time. For those of you who have never heard of "poor man's fertilizer," or who might not know to reach for an umbrella if you find bubbles in your cup of tea, I hope this book triggers a new curiosity about our connection with nature.

My connection to weather, my passion for what I do every day, comes from my family and from an incredible childhood on the farm—so many little things that today mean so much to me. I remember waking up in the morning, hopeful that Dad would take my brother Ronnie and me fishing. Before Dad was in from the barn, our trip to the river was cancelled because Grandma had stepped out onto the veranda and come back in to proclaim, "The wind is from the east, the fish won't bite today." We were disappointed, but we knew better than to question Grandma or Mother Nature. The weather was very much a part of farm life. When the weather was fair, we worked in the fields; when the rains came, we worked inside and tried to predict when the weather would clear.

I feel so blessed to have been raised on a farm. There was a time when my friends who lived in town used to joke about how dull it must be to live way out in the boonies. For a very brief time, I envied their lifestyle. They could hang out on the post office steps or meet for a soda at the fifteen-cent store. That envy was fleeting. Life in the country was anything but dull. It was busy, challenging, exhausting, but above all, quiet and educational. Nowhere else could learning be so much fun. After all, nature is an exciting, living classroom; explore it, love it, learn from it, but above all, respect it.

Grandma Says

When sparks fly, a storm is nearby.

• •

Y EARS AGO, ALMOST every home had a wood stove. Today, we have electric fireplaces to set the mood, and gas fireplaces that we can turn on with a switch. Times have changed. But there are still some wood-burning stoves out there. If you have one, you've probably noticed that some days, very few sparks fly when you open the door to stoke the fire, but other days, sparks fill the air. Well, you've guessed it. Grandma would tell you that was a sign of incoming rain. When I started to study meteorology and the physics of the atmosphere, I couldn't wait to find out why this was. Well, as is the case with many of Grandma's sayings, there's some pretty solid science behind it.

Air is a mixture of gases, largely composed of nitrogen and oxygen. It's generally considered an insulator, and would be an excellent one if all the oxygen and nitrogen molecules were in a neutral state. However, the air is actually composed of varying quantities of neutral molecules and positive and negative ions. As the number of ions in the air increases, the air progressively becomes a better conductor. So where do these ions come from? Precipitation! High in the upper atmosphere, you'll find large, less mobile ions. Falling rain, snow, or even ice ahead of a weather system tends to pull these ions down toward the earth, making the air a better conductor. So "when sparks begin to fly, a storm is nearby."

WINTER

If you're wondering how many times it will snow during the winter, you'll need to find a cat.

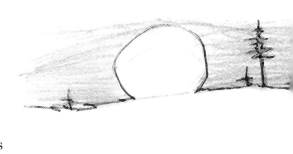

THE MOON HAS been mystifying humans since the beginning of time. Centuries ago, we turned to the moon's twenty-nine-day cycle to set up many of the calendars still in use around the world today. Without the moon's cycle we would have had to develop an entirely different way to keep track of time.

I wouldn't say that Grandma had lots of time, but she made time for the moon. If the moon was out, Grandma had something to say about it. She would comment on its phase, its position in the sky, and how clear it was or wasn't. Earth's only natural satellite also helped Grandma predict the weather, or at least the number of snowfalls we could expect over the next few months. In late fall, when the air cooled just enough to bring a threat of flurries, Grandma kept an eye on the ground; she was waiting for the first snow cover of the season. A light dusting wasn't quite sufficient; there had to be enough snow on the ground to be able to see a cat's tracks. The timing of that snowfall would help Grandma calculate the number of times it would snow during the upcoming winter. Grandma believed that if you took the date of the first snowfall in which you could see a cat's tracks and added it to the age of the moon, you would get the number of snowfalls for your area. Figuring out the age of the moon is not difficult. The new moon is your starting point. The day after the new moon, the moon is one day old. The full moon is halfway through the twenty-nine-day cycle; the day before the next new moon, the moon is twenty-nine days old. Now this doesn't tell you how much snow will fall, just how many times it will snow. I've done this for several years and it always comes surprisingly close. Try it. I can't say much about its connection to meteorology, but it's lots of fun!

The colder the snow, the louder the crunch.

I REMEMBER THE FIRST time Grandma shared this little weather gem with me. We were walking back from the barn on a clear and very cold night. The stars were dancing overhead and it was so cold you could see your breath. It was one of those amazingly quiet winter nights, except for the squeaking sounds Grandma and I made with every step. After a dozen steps or so, Grandma turned to me and said "it's a cold one, must be about minus eighteen." I didn't say anything, but quickly ran over to the side of the shed where Dad had hung a lovely thermometer. She was wrong—by one degree! When I asked her how she knew, she said that the colder it was, the higher the squeaky pitch would be when you stepped down on the snow.

I've come to learn that she was absolutely right. The air temperature and the quality of the snow are both factors that determine if snow will be noisy or quiet underfoot. Snow has air trapped between each flake, and when stepped on, those air spaces absorb sound. Dry, fluffy, new snow has more air trapped between each flake, resulting in quiet footsteps. Hardened, old snow has less air trapped between each flake, which means that less sound is absorbed, resulting in noisy or squeaky snow. While this is true, it only really works on very cold nights. Snow only makes sound when the thermometer dips below minus ten degrees Celsius. Temperatures above that allow the snow to melt just enough to slip silently under your boots as you walk. So your boots can be a good indicator of just how cold it is outside in the winter.

After many cold walks to the barn, Grandma concluded that if your steps are making a "crunch, crunch, crunch" sound the temperature sits somewhere between minus ten and minus eighteen. When you hear that distinctive "squeak, squeak, squeak," bundle up—it's at least minus nineteen degrees. I hope your next walk in the snow is spent listening and learning. That's what Grandma did best.

If it snows on Christmas night, the crops will be good.

G RANDMA WASN'T THE only weather prognosticator in the family. Many of my relatives were from the country and had become quite proficient at short-term forecasting. When family got together for Christmas, it was a good time for the farmers in the family to go over the fall harvest and recount stories of the late spring frosts, the thunderstorms that flooded the back fields, and the late summer drought that threatened a second cut of hay. That usually took us to mealtime. After the feast, the storytelling started up again; stories about how my grandfather would note the hours of sunshine on Christmas Day so that he would know how many times frost would come in May. An uncle of mine remembered his father telling him that if it snowed on Christmas night, the crops would do very well the following year. That reminded a cousin that his mother believed that if ice hangs on the willow tree at Christmas, the clover would be ripe for cutting at Easter.

There were many weather observation made between Christmas and January 5. Our ancestors referred to these twelve days as the "Ruling Days," as they believed that they ruled the weather for the year ahead. It's a different twist on the twelve days of Christmas; there's no singing involved, but a little homework. You'll need a pad of paper and a pencil to keep track of things; I prefer to use a calendar—you'll see why in a moment. The "twelve days of Christmas" is the time between December 25 and January 5; those dates used to be known as "new" and "old" Christmas, respectively. The kind of weather you receive on those twelve days is believed by many to point to the weather you'll have in the year ahead. For example, if a cold rain falls on December 25 then January will be cold with lots of rain. If December 26 is sunny, then February should be a bright month. The weather on December 27 will tell you what the third month, March, will be like. I'll stop there, but you can see how this works, and why jotting down the weather on a calendar is very important. It's something fun you can do with the kids throughout the holidays, but the real fun begins as you monitor the accuracy of "the twelve days of Christmas."

If you hear thunder during Christmas week, it's going to be a very snowy winter.

• •

THERE ALWAYS SEEMED to be a lot of weather watching going on at the farm around Christmas. That's the time winter usually begins to settle in, so it makes sense that our ancestors were looking for signs of the season. I've compiled a few of Grandma's favourite Christmastime weather sayings. While they're very interesting, they really can't be explained scientifically, but have been observed and talked about for generations; that has to count for something. I remember Grandma saying that if we heard thunder during the Christmas week, it would be a very snowy winter. That always made me quite happy, and wishful that Santa would place a new Crazy Carpet under our tree. While Christmas morning was always very chaotic, as soon as the sun was up Grandma would start her Christmas Day count. She believed that the hours of sun on Christmas Day would equal the number of frosts in May. Now wouldn't it be fun to keep track of that one? Let me know what you come up with.

On a less cheery note, the most morbid Christmas weather saying has to be, "A green Christmas makes a full graveyard." Actually, Grandma would say that one in French: *Un Noël vert rempli le cimetière*. This grim prediction refers to the belief that an unusually mild winter causes more disease. Grandma certainly was among those who believe that cold weather is healthier. According to weather lore, if the ground is frozen and covered in snow, germs are less likely to spread; if the grass is still green on Christmas Day, germs and disease will spread more readily. Today we know that cold doesn't kill germs, the germs simply become inactive when it is cold. Grandma used this little bit of folklore to her advantage. She used to tell the kids that we should run around outside to clear our lungs of the germs spread by all the people in the house. Grandma was pretty wise. I wonder if she really believed that or if she just wanted a little peace and quiet. To this day I still believe that a good blast of fresh, cold air is very healthy.

WINTER

Little flakes, big snow; big flakes, little snow.

A H, WINTER. I love winter. Every year I'm amazed at the number of people who wish the season away. They think it might not come if they don't wear boots and a winter coat; they walk around looking for the elusive robin, and they count the sleeps until the groundhog makes his prognostication. I don't get it. This is Canada!

I think the prettiest thing about winter is the snow. The flakes are tiny little masterpieces. I love to catch them on my tongue. When I was young, I couldn't get my snowsuit on fast enough when those dainty flakes started to fall. It was

always interesting to hear what Grandma had to say about the snowflakes. You've no doubt heard the expression, "Little snow big snow, big snow little snow." It implies that the smaller the snowflake, the more significant the snowfall will be. Conversely, the larger the snowflake, the less shovelling you'll have to do. So what's going on up there?

The size of the snowflake depends on moisture and wind or turbulence. The fabulous snowflakes that look like crochet doilies fall from convective type clouds, the kinds that produce summertime showers. Updrafts and downdrafts found inside a convective cloud enable the snowflakes to grow to such an impressive size. The updrafts keep the flakes in the cloud until they become large and heavy enough to fall to the ground. In the meantime, as they bounce around inside the cloud, they stick to each other, forming large, intricate flakes of snow. These impressive flakes fall in the form of flurries. By nature, flurries start and stop and don't usually last very long, therefore snowfall amounts are relatively light.

The smaller flakes usually fall from a more stable type of cloud; the water vapour freezes and forms tiny snowflakes that fall to the ground. Did you ever walk outside just before the snow started to fall and notice how grey and uniform the sky looked? That's a classic pre-snow cloud called a nimbostratus. It's a large, sprawling cloud that accompanies a well-developed weather system. It often serves up hours and hours of snow—the tiny flakes that end up giving you a large pain in the back.

If the groundhog sees his shadow, we're in for six more weeks of winter.

GROUNDHOG DAY. GRANDMA was a big fan. What's not to love? Kids gather, cameras flash, and hot chocolate is served. I have to be honest with you; it's not my favourite day. Think about it. I work very hard every day analyzing charts and building maps to prepare the most accurate forecast possible. This rascally rodent sleeps for a few months, wakes up and emerges from his lair for air on February 2, and everybody is watching and talking about him! I guess it's a little professional jealousy.

Despite all of that, Groundhog Day is without a doubt the most talked about North American weather folklore. If it's cloudy when the groundhog emerges from its burrow on February 2, it will leave the burrow, indicating an early spring. If however, it's a sunny day, the groundhog will supposedly "see its shadow" and scurry back into its burrow, and winter will continue for six more weeks.

The tradition dates back to 1887, and though the origins are unclear, it is said to have originated from ancient European weather lore in which a badger predicts the weather. It also has connections to an ancient celebration called Candlemas, a point midway between the winter solstice and the vernal equinox. Superstition has it that fair weather on Candlemas was seen as a prediction of a stormy and cold second half to winter.

Back to the badger or groundhog. Have you ever wondered how accurate the furry rodent is? Let's just say he's not batting a thousand. In fact, since 1988, Shubenacadie Sam was "right" ten times and "wrong" thirteen times. With an accuracy rate of forty-three per cent, the little fella better not quit his day job, whatever that is!

When the chimney smoke descends, our nice weather ends.

WINTER IS A magical season. On a clear, cold night millions of stars dance overhead, while snow squeaks underfoot and the smell of chimney smoke fills the air. Most nights, that smoke goes straight up, but that's not always the case. Some nights you'll find that the smell of smoke wafting through the air is much more pungent, and if you look closely, you'll see that the smoke is hovering overhead. You can blame that on the weather. Grandma used to say, "When the chimney smoke descends, our nice weather ends." She used to point out that when smoke from the chimney sank to the roof then fell to the ground, there would be snow or rain within twenty-four hours.

She was right. When a storm is approaching, there is more moisture in the air. The dirty particles in rising smoke absorb that added moisture. This makes the smoke heavier and instead of rising freely, it's dragged down. Conversely, in dry, calm weather, a column of hot smoke from a chimney rises straight up and will continue to rise as long as it is warmer than the surrounding air. So smoke that rises vertically is a sign of fair weather. The next time you're out for a stroll, enjoying the smell of wood smoke in the air, don't forget the lesson Mother Nature is trying to teach us.

Ground fog in the summer is a sign of fair weather; ground fog in the winter forecasts rain.

I T'S USED IN movies to thicken the plot and in spas to clear your skin. Fog. Grandma used the fog to—you guessed it—predict the weather. Many old-timers would tell you that "ground fog in the summer is a sign of fair weather, whereas ground fog in the winter forecasts rain." Grandma wasn't a scientist, but decades of observation had proven this to be very true. When I started to study meteorology, I was pleased to learn that the connection between ground fog and the weather could be explained.

During the summer, morning ground fog is a result of overnight cooling. It's called radiation fog and it takes place on a clear, windless night. Clear skies and light winds occur when there's a fair weather system directly overhead. If the fair weather system was overhead overnight, it would take at least the full day before it moved out. You could therefore count on a nice day ahead.

Ground fog in the winter is a result of warm air moving in ahead of a front. That fog is known as advection fog. It forms when the moist, warm air mass slides over frozen or snow-covered ground. A winter warm front usually brings rain. So the next time you're enjoying the beauty of a fog-draped landscape, you can predict the weather for the day ahead.

Nova Scotia's Three Snows

ONE SATURDAY MORNING at the local farmers' market I met a lovely man who had a lot to say about the weather. Like Grandma, he had spent decades observing nature's little quirks. He had heard me mention the sugar snow and asked if I knew about the Valley's "three snows."

According to Maritime folklore, the Annapolis Valley always receives three snowfalls after the March equinox. They are so predictable that they've been given names: smelt snow, robin snow, and grass snow. Local fishermen know that the smelt start to run right after the first spring snow. The robin snow ushers in the spring robins. (Many North American robins don't migrate, but flock to more remote areas for shelter through the winter. According to folklore, the second spring snow brings them back.) The grass snow, welcomed by gardeners, is often referred to as "poor man's fertilizer."

Can these "snows" be explained? Well, in the spring, as weather patterns begin to shift, warm frontal systems bring a return to moderate conditions, and big, fluffy snowflakes! Behind a March warm front, a combination of sunshine and warm winds will push temperatures above freezing. That weather trend and temperature profile could trigger smelt migration, bring birds out of the woods, and maybe even green up the grass. It's not uncommon to get several snowfalls after the first day of spring. Just for fun, I turned to the record books and found that we've had three snows following the vernal equinox in eight of the past ten years. So while many believe that the vernal equinox marks the arrival of spring, our ancestors knew that spring hadn't really set in until after the "three snows."

Sugar snow makes the sap flow.

IS THERE ANYTHING simpler yet more delicious than maple syrup? I'd have to say that it's one of the many wonders of the world. Around here, it's an undisputed sign of spring. As the days start to warm and the nighttime temperatures fall below freezing, the sweet sap begins to work its way up from the roots. Depending on the weather—and that varies from year to year— the maple trees are tapped anytime between late February and mid-March. Knowing when to tap can be tricky, unless you're as wise as Grandma. Each spring when the topic came up at the dinner table, Grandma was quick to mention the sugar snow: "Sugar snow makes the sap flow."

With the first signs of spring, Grandma watched carefully for a light fall of large snowflakes. She believed that the sweet sap would start to flow after that spring snowfall. In the spring, when big puffy snowflakes fall and the accumulation is fairly light, it's often the result of the passage of a warm front. Behind a March warm front, a combination of sunshine and southwesterly winds will push temperatures above freezing; because it's March, temperatures fall back after the sun sets. Warm days followed by below-freezing nights will fill the sap buckets pretty quickly. And of course, after boiling down the sap, you're left with beautiful, golden maple syrup.

I have to tell you about something my very clever—not to mention thoughtful—mother did every year. Before all the snow had melted away, Mom would fill an old turkey roaster with snow and put it in the freezer. When we least expected it, usually on a hot summer day, she'd take the snow out of the freezer and we'd have taffy on the snow. A truly magical childhood memory! Tasty, and very Canadian.

If the frogs are singing, there's rain coming.

SOMETIMES GRANDMA WOULD talk about being able to "smell" the rain coming; other times she could "hear" it. One night, we had finished our barn chores and were walking back to the house when Grandma said, "Listen to the frogs. Can you tell what they're saying?" The frogs were very loud that night. Their melodious trill filled the air. Frogs were very mysterious to me. We didn't see them very often, but some nights we couldn't miss their song. I didn't know what the frogs were saying that night, but Grandma sure did. She explained that when the frogs were especially loud, it would soon rain.

Grandma was absolutely right. Frogs are associated with weather predictions because they begin croaking just before it rains. Frogs prefer humid weather, and will sing when the air is humid: of course the relative humidity rises before it rains. I guess a damp frog is a happy frog, and that makes it sing!

Scientists say that frogs can tell us a lot about the weather in general. Their skin is extremely thin and sensitive, and they respond to even small changes in atmospheric moisture and temperature. Scientists believe an analysis of the frogs' song, combined with readings from climate data, could help improve our understanding of the impact of climate change—and of frogs.

And lastly (but please don't try this at home), according to European folklore, frogs kept in a glass were able to forecast the weather. People would put some water in the glass to keep the amphibian happy, and then add a small ladder. A climbing frog would indicate good weather, whereas a frog hanging out in the water would indicate bad weather. This belief especially stuck with people in the German-speaking countries where weather forecasters are typically called "weather frogs." I hope that doesn't catch on here!

Oak before ash, only a splash.

· ·

WHILE CRAZY, OUT–of-order temperature swings in the spring can't be trusted, Grandma believed that the oak and ash trees could be. According to popular weather lore, the order in which these trees bud is a sign of the amount of rain we can expect in the coming season. It goes like this: "Oak before ash, only a splash; ash before oak, we're in for a soak." If the oak buds appear first, the summer will be relatively dry. If the ash buds appear first, we're in for a wet summer.

Grandma wasn't the first to watch for this. Scientists at the Centre for Hydrology and Ecology in England have records dating back to the eighteenth century. According to their data, the race between oak and ash was a far more equal one in the last century; then the oak came out ahead about 60 per cent of the time. It now appears that climate change has made the competition between ash and oak a very unequal one. Over the past couple of decades, oak has won ninety per cent of the time. The ash tree has shallow roots and relies on moisture from winter's snow cover to set its buds, and we're not getting nearly the same amount of snow as we did thirty years ago. On the other hand, the stately oak does much better in drier conditions, or without the spring snowmelt. This could be yet another sign of how some of our native habitats could be altered as climate change takes hold.

A cold May fills the barn with hay.

· ·

AS MUCH AS I love winter and of course snow, in due time I do get pretty excited about the return of spring. Now, I should point out that spring in the city is quite different from spring in the country; in rural areas it takes a little longer to settle in. While the heat from the sun is being soaked up by the concrete and asphalt in towns and cities, the sun's rays bounce off the ice and snow on the fields in the country, and the winds blow cold. When I was young, there were some days it looked like we country kids were coming to town from another planet. I remember getting off the school bus bundled up in my winter coat, only to be greeted by my friends wearing cute little sweaters. And while paved roads were dry and the town girls wore their fancy shoes, we were still in our boots. You see, early spring wasn't always beautiful on the farm. After a hard, cold, snowy winter, things turned to mud for awhile. So it's fair to say by the time May arrived, we were ready for some heat.

Grandma, on the other hand, was never in a hurry for the warm weather to move in. If we complained about the cool weather dragging its heels in May, Grandma was quick to remind us that "A cold May fills the barn with hay." That expression has been around for a very long time. Perhaps it was more valid in the days before all the fertilizer use, back when we allowed plants to evolve a little more naturally, but the thinking behind this weather saying is pretty solid. A cool, damp spring encourages sturdy, if not slow, crop growth and discourages early insect pests. These strong plants could produce better harvests, leaving more to store through the following winter. So aside from dressing a little differently in the spring, in some ways, our ancestors' wisdom and their respect for nature made country kids think a little differently too.

A robin is a sure sign that spring had arrived.

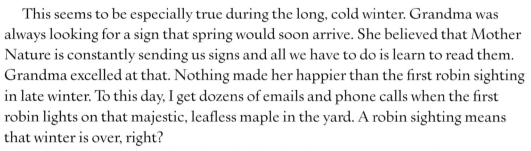

WE'RE OFTEN REMINDED that we should "live for the moment." Despite that very good advice, we always seem to be wondering what's ahead. I guess that's human nature.

This seems to be especially true during the long, cold winter. Grandma was always looking for a sign that spring would soon arrive. She believed that Mother Nature is constantly sending us signs and all we have to do is learn to read them. Grandma excelled at that. Nothing made her happier than the first robin sighting in late winter. To this day, I get dozens of emails and phone calls when the first robin lights on that majestic, leafless maple in the yard. A robin sighting means that winter is over, right?

Well, not exactly. The American Robin is migratory. However, the robins we have in Nova Scotia are not "complete" migrants. That means they don't all leave their breeding range during non-breeding months. Here in the East, many robins stay put during the cold months. They're more likely to stick around in areas that provide a reliable supply of food. We see fewer robins in winter because they are less conspicuous. They spend most of their time in flocks and prefer rural areas where there are fewer people, but available food. When spring arrives, the robins become more obvious as they move back into more populated areas and the flocks break up. They call this activity the breeding dispersal. So, while a robin sighting is a lovely thing, it doesn't necessarily mean that spring is here.

If the tap water is cold, there's a mild spell coming.

I LOVE THE taste of well water. During a recent trip home I was surprised by a comment my dad made. He had just gone to the kitchen tap for a glass of water and said, "Wow, that water is cold; we're in for some milder weather." This was something I had heard Grandma say a few times, but after moving to the city, I had forgotten all about it.

We have to go underground to explain this one. The temperature of water from wells is remarkably constant. In wells that are from ten- to twenty-metres deep, the water temperature is one or two degrees Celsius above the annual mean temperature of the area. Soil temperature varies slightly from month to month as a function of solar radiation, rainfall, seasonal swings in air temperature, ground cover, type of soil, and depth in the earth. The fluctuations in soil temperature, especially ten or more metres below ground, are much less dramatic than the changes in air temperature. That's because soil has a much higher heat capacity than air, meaning it holds the heat for a longer period of time, and releases that heat much more slowly. For that reason, seasonal temperature changes in the ground lag behind seasonal changes in air temperature. For example, the temperature of your well water will drop in the winter but not until the cold has settled in for a few months.

Back to the kitchen tap. It was usually sometime in March when Grandma would comment on the very cold water from the well. It was a great observation. It had taken all winter for the well-water temperature to drop a couple degrees. Since it was March, we were likely to get some milder weather soon. So while this weather saying is really more of an observation, it's a bit of a science lesson, too.

Wherever the wind lies on Ash Wednesday, it will continue in that quarter during all of Lent.

I GREW UP in a Catholic household. Some Christian holidays or feast days were more involved than others. In most cases, there was some personal preparation followed by some time spent at church. For me, the least anticipated of them all was Ash Wednesday. Ash Wednesday is the first day of Lent—the forty-day period leading up to Easter. It's a time of penance. I've always given something up for Lent; usually chocolate because it's what I love most. Grandma was quick to remind us that a little penance never hurt anyone. But penance wasn't the only thing that Grandma observed on Ash Wednesday; she also checked the wind. According to Grandma, "Wherever the wind lies on Ash Wednesday, it will continue in that quarter during all of Lent."

Some weather sayings are backed by scientific reasoning; this one is a little more obscure. It could be explained by seasonal change. In the spring, the infamous jet stream, that controlling band of wind in the upper atmosphere, starts its move. It shifts from a typical, more southern position in the winter to a more northerly placement for the summer. Decades of observation seem to imply that if that shift has not started by Ash Wednesday, it won't take place until after Easter. Therefore, the wind direction observed on Ash Wednesday will remain for forty more days, the duration of Lent. In the grand scheme of things, forty days is far from an eternity, but if the wind is from the north and one can't eat chocolate, it certainly feels like it!

As the days lengthen, the cold strengthens.

I LOVE WINTER, so I'm not one of "those" people who count down the days until spring. Doing that can be disappointing. How many times have we made it to the end of February relatively unscathed only to have our hearts broken by March? Not only can we get our worst storms late in the season, but we can experience some pretty cold temperatures as well. Many people falsely believe that once the days start to grow longer, the temperature begins to warm. Grandma knew better. She used to say, "As the days lengthen, the cold strengthens."

Grandma had observed this to be true; it's also scientifically correct. But why? After all, it's on or around December 21 that we receive the least amount of daylight. As the daylight minutes grow, shouldn't it get warmer? Not really, at least not right away. The reason is "seasonal lag," the phenomenon whereby the date of minimum average air temperature at a geographical location on the planet is delayed until sometime after the date of minimum insolation or daylight. This also applies in the summer when the maximum average air temperature is delayed until sometime after the longest day of the year.

Earth's seasonal lag is largely caused by the presence of large amounts of water. Basically, it takes longer to warm those large bodies of water, and of course it takes longer for them to cool too. This difference is not "seasonally symmetric," meaning the time between the winter solstice and coldest period is not the same as between the summer solstice and hottest time. In mid-latitude continental climates, it is approximately 20–25 days in winter and 25–35 days in summer. That would put our coldest period at about mid-January, almost a month after the shortest day.

Planting by the light of the moon

GRANDMA ALSO KNEW that when the songbirds returned and the sap was flowing in the sugar bush, it was time to start thinking about the garden. Our garden was incredible. Okay, to be honest it wasn't weed-free, but the produce that came off that plot of land was amazing. I'm pretty sure the abundant harvest had a lot to do with our rich farmyard compost and Grandma's knowledge of when to plant the various seeds.

Grandma always started her tomato seeds on Good Friday. You're probably wondering, "What does that have to do with the moon?" Everything. People know that the date of Easter varies every year, but many don't realize that the

date is set according to the moon. As I mentioned earlier, Easter is celebrated on the first Sunday following the first full moon following the vernal equinox. Like many gardeners, Grandma planned her planting schedule by the phases of the moon. Gardening by the moon is called lunar gardening and it doesn't mean planting at night, but understanding which plants benefit from being planted during a particular moon phase.

Every month there are four phases, or quarters, of the moon's cycle, with each phase lasting about seven days. The first two phases, when light increases from the new moon to the full moon, are referred to as waxing. The two phases after the full moon, when light gradually diminishes, are called waning. The four moon phases influence the earth's gravity, which is why the ocean tides change and groundwater rises and falls. According to old-time gardeners, the waxing phase—the second quarter—is the best lunar phase for planting tomato seeds. The waxing moon encourages plant growth because the groundwater continues to increase as the moon moves toward its full phase. For the very best results, you should try to plant your tomato seeds two days before the full moon—Grandma aimed for Good Friday. We always had gorgeous plump tomatoes by late summer, not to mention something to do after church on Good Friday.

Fish don't bite on an east wind.

GRANDMA LOVED FRESH fish. She didn't like to go fishing, but she didn't mind cooking up whatever we brought home. Farmers don't get to do a lot of fishing; most times when the weather is nice, they're in the field. That's why it was so important to make the most of the best fishing days. Grandma kept a close eye on the wind direction. In the morning, she would let us know if we should bother heading out. She believed that the fish wouldn't bite if the wind was from the east. A few— very few—times, we ventured out for the day despite Grandma's warning. We also came back empty-handed.

So what is it about the east wind?

The wind turns to the east ahead of a low pressure system. The belief behind this lore is that when air pressure lowers, gasses created by decaying plant matter, which resides on the bottom of the body of water, begin to release. Bubbles containing the tiny micro-organisms that live in these plants rise to the surface, creating a feeding frenzy among the fish. With so much food around, the fish are not hungry enough to bother with bait. So before you head out fishing, check your tackle box, your bait bucket, and above all, check the weather vane.

If it rains on St. Medard, it will rain for forty more.

J UNE IS A pretty month. Flowers come into bloom, the air warms nicely. As soon as the calendar page flipped over to June, Grandma started to look ahead to the eighth. June 8 is St. Medard Day. According to Grandma, "If it rains on St. Medard, it will rain for forty more."

Over the years, I've wondered if that weather saying was widely known. Last year, I got an email from Emerise Germain who was as pleased as could be that the sun was shining brightly over Meteghan Station on the feast day of St. Medard.

So who was Saint Medard? He was born into French nobility in the fifth century. He was a pious and excellent student and was ordained to the priesthood at the age of thirty-three, and then became a missionary bishop who used his personal wealth to establish a scholarship fund. While still a youth, he gave one of his father's finest horses to a peasant who had lost his. Immediately afterward, there was a sudden downpour, and while everyone else was drenched, an eagle spread its wings over Saint Medard, who remained dry. Saint Medard's feast day is June 8, and the French say, "*S'il pleut le jour de Saint-Médard, il pleut quarante jours plus tard*." But if it is sunny and dry on Saint Medard's Day, the next forty days will also be dry.

I wasn't able to find any weather statistics to either support or debunk this one, but perhaps you could start tracking it. I know a lot of people have made the move to electronic calendars, but there's something to be said for an old-fashioned wall calendar. You can jot down notes, compare weather from year to year, and draw some of your own weather conclusions. You should have seen Grandma's calendar!

If it rains on St-Jean Baptiste Day, expect a wet harvest.

G RANDMA'S CALENDAR WAS really something. I don't know how she kept track of things; she had so many dates circled. On the June calendar page, the twenty-fourth was one of those dates. On that date, many people around the world celebrate St-Jean Baptiste Day. In France, the origin of the holiday was the pagan celebration of the summer solstice—a celebration of light and a symbol of hope. The holiday became Christianized and the French who settled in North America in the 1600s continued to celebrate the event. The annual celebration grew and in 1925 the Quebec legislature declared June 24 a holiday.

Before Grandma went to bed on the twenty-third, she would go outside and hang her prayer beads on the clothesline to keep the rain away. According to weather lore, if it rains on St-Jean Baptiste Day, you could expect a wet harvest.

June 24 is also known as midsummer, which simply refers to the period of time centred on the summer solstice. Some people believe that Midsummer Day dew has special healing powers. Apparently, it's good to walk barefoot in dew on the morning of Midsummer Day; it saves the skin from getting chapped. And here's my favourite: young girls wash their faces with midsummer dew to make themselves beautiful.

So I suppose to keep farmers and young maidens happy, we need a clear morning with a heavy dew on St-Jean Baptiste Day.

In like a lion, out like a lamb.

A FTER "RED SKY at night," "In like a lion, out like a lamb" is without a doubt the most-recited weather proverb. Grandma loved it. She looked forward to the first day of March for weeks.

This weather saying goes back centuries and is a little different from the others. Most weather lore is based on careful observation of nature's cycles or of animal behaviour. Well, there were no lions on the farm, but there was a famous lion high above the old grey barn: Leo the Lion.

According to astronomers, this phrase may have originated from ancient observations of the stars. In the springtime, two of the constellations visible just after the sun has gone down are Leo, "the lion" and Aries, "the ram" or "the lamb." Leo rises in the east at the same time that Aries sets in the west. While this happens throughout the year at varying times, it's most visible in early night sky of March. If the sky was clear on March 1 and Leo was spotted overhead, it was believed that since the lion had ushered in the month, it would end on a quiet note. Conversely, if clouds covered the night sky on the first day of March then the lion would roar at the end of the month, and the weather would be stormy. But has the popular phrase held true to its weather predictions? A number of years ago I decided to check it out. I found that over a ten-year period spanning from 2000–2010, the popular weather saying was right six times. The proverb might not always be right, but after a long winter it can serve an important purpose: it might bring some hope.

Beware of Sheila's Brush

I WAS RAISED in a French Catholic home; it's pretty safe to say that there's not a lot of Irish blood coursing through my veins. Despite that, each year as March rolled on, Grandma was on the lookout for Sheila, more specifically, Sheila's Brush. As a child I really enjoyed all of Grandma's wonderful weather wisdom, but I loved to hear some of the stories that went along with them. So let's get back to Sheila—who was she? Well, according to Grandma, and popular Newfoundland legend, Sheila was the woman in Saint Patrick's life. Depending on which historian you believe, she was his sister, wife, mother, mistress, or even his housekeeper.

Here's how Grandma told it. By mid-March, Sheila, like many of us, was fed up with winter and looked forward to spring. One year, when snow started to fall on or around March 17, she got out her broom or "brush" to quickly sweep it away. Her vigorous sweeping stirred things up and caused a storm. So in her effort to brush winter away, Sheila triggered one last snowstorm for the season.

There are many examples of Sheila's Brush. On St. Patrick's Day, 2008, the second of two powerful back-to-back storms roared across Newfoundland. Schools and businesses were shut down. In St. John's, even public transit was pulled off the road. Roads were completely blocked by snow. Gander saw 120 centimetres of snow—about a quarter of its average annual snowfall—in about a week!

Prayer beads on the clothesline
will keep rain off the bride's head.

· ·

MY SISTER MONIQUE GOT married in September. A fall wedding can be perfect—not to hot, not too cold, hopefully not too wet. Most brides can cope with a hot day or even an unseasonably cool one, but rain, now that just can't happen.

As my sister's wedding day approached, she started to fret over the forecast. She checked the weather several times a day. A week out, things looked good, then the forecast changed and Monique's anxiety level soared. Grandma didn't understand my sister's emotional roller coaster. There was nothing she could do—or was there?

Well, it became apparent that Grandma was pretty cool about it because she had a plan. She told Monique to go up to her bedroom and get the prayer beads she had received for her first communion. She came down with them, but told Grandma that she had a million things to do and didn't have time to pray. Well, that wasn't the point of getting her beads. Grandma then told us that for decades, brides would hang their prayer beads on the clothesline the night before their wedding day to keep the rain away. Of course Monique did, and it worked!

Now does this always work? Grandma says that it does…if you believe. This is one of those cases where the big guy might trump science.

If it does rain on your wedding day, all is not lost. According to folklore, rain on your wedding day is a sign of money to come. Armed with this valuable information, I wonder how many brides might prefer a little wet weather on their wedding day?

When March blows its horn,
your barn will be filled with hay and corn.

W HEN IT COMES to forecasting the weather, no month is more challenging than March. It's not uncommon for a system to move through with some freezing rain and ice pellets ahead of it, rain behind it, and thundershowers triggered by the cold landmass below.

Those March thunderstorms always got Grandma's attention. She used to say that "when March blows its horn, your barn will be filled with hay and corn." "Blows its horn" refers to thunderstorms. As a former tuba player, that always makes me chuckle. So can a March thunderstorm be an indicator of a good crop down the road? I have to admit this one is a bit of a stretch. March thunderstorms indicate unusually warm weather. Thunderstorms occur when you have a significant temperature difference between ground and sky and sufficient amounts of moisture to produce charge differential inside the cloud; it's not really an indicator of the long-term weather trend. (The charge differential occurs when the positive charges or protons rise to the top of the thunderstorm cloud while the negative charges or electrons sink to the bottom. Since thunderstorms are localized over a small area, they are not likely to hold the key to a long-term forecast.)

Having said that, enough warm air so early in the season could be a sign that the jet stream is on the move, transitioning from its southern winter position to a more northern location for the warmer months. If farmers are able to get out on the fields and work the land early, the crops will be in with lots of time to mature.

It's not snow, it's poor man's fertilizer.

· ·

W HILE MOST PEOPLE cringed at the thought of a late spring snow, Grandma rejoiced. She called it "poor man's fertilizer." Grandma believed that this snow would somehow be good for the lawns and gardens. I had heard that snow contained nitrogen, but had always been a bit skeptical.

Well, it turns out that snow can be beneficial, and more so now than ever before. The levels of nitrogen, sulphur, and other elements in snow have increased over the last several decades. Most of the nitrogen comes from emissions as a result of industrial manufacturing and burning fossil fuels; the rest comes from lightning. While the extra nitrogen can be a problem in terms of the acidification of soils, the soil that we use for gardens and lawns could usually use a little nitrogen boost. The bottom line is, snow contains nutrients and a lot of moisture. And if that snow falls on ground that's not frozen, as would be the case in late spring, then the nutrients and moisture in that snow can penetrate into the soil and actually do some good for the plants that will grow in it later in the year. So while we might feel cursed when that late spring snow flies, we'll be blessed with lush, green lawns come summer.

If the dandelions stay closed after sunrise, there's rain on the way.

MY SISTER AND I dreaded the first mild stretch of weather in May. That's when Grandma would inform us, "This is the weekend that we're going to pull dandelions." It really wasn't that bad on Saturday, but by the time Sunday morning rolled around, we'd had enough. I remember walking over to my bedroom window early one Sunday to assess the job at hand, only to see a carpet of green; not a dandelion in sight! I thought an angel must have come overnight to spare my sister and me. By the time we got dressed and made our way to the breakfast table, the news got even better. Grandma told us that we wouldn't be pulling dandelions because there was rain coming. According to Grandma, if the dandelion flower stays closed after sunrise, there is wet weather on the way.

Different flowers open and close for a variety of reasons. Gardeners have observed this for centuries and scientists have looked into it too. Many researchers, including Charles Darwin, have speculated that flowers may have evolved certain traits or structures to protect themselves against the damaging effects of rain, which can wash away pollen grains and dilute nectar. The dandelion is one of those flowers. Another, more popular, example is the lovely scarlet pimpernel, which has been called the "poor man's weather glass." Its flowers open in sunny weather, but close tightly when rain is expected. The petals of the morning glory act in a similar way, with wide open blooms indicating fine weather, and shut petals predicting rain and bad weather.

Opening and closure in other species is regulated by changes in light. Take the tulip for example. Its petals open in the morning and close in the evening. In other species you'll see a nocturnal opening; this is triggered by a rise in relative humidity. The moonflower is one of those unique evening beauties. Once the sun goes down, the big white flowers spring into action and open their petals. If you plant them on a trellis, it looks like they're reaching for the moon. Sometimes a short-term forecast is as easy as a quiet trip to the flower beds.

Flies bite before it rains.

· ·

FLIES. WHAT'S SUMMER without them? We had more than our share of them at the farm…for obvious reasons. They drove Mom crazy. The "fly specs" on the white house were a pet peeve of hers, and the ones on the white car made my dad mad, but Grandma had a different kind of appreciation for the common fly. Flies helped Grandma hone her forecast skills. She believed that if the flies were biting, it was going to rain before long.

It seems a little odd, but there is a scientific connection. Before it starts to rain, the relative humidity of the air rises. Flies, in fact many insects, are more likely to settle on objects during moist weather; they find it quite difficult to fly under these conditions. If that's not enough to convince you, I'm going to throw in the appealing topic of body odour. When atmospheric pressure decreases, we tend to perspire more easily. While this release of sweat may not appeal to you, it makes us a more appetizing target to flies. For these reasons, insects become more bothersome just before it rains.

No thermometer?
Count cricket chirps instead.

B ACK ON THE farm, my bedroom window was always open and I welcomed in all the sounds that nature served up. I especially enjoyed the crickets. Grandma's bedroom was on the same side of the house, so we enjoyed the concerts together. I was very young when Grandma explained to me that those cute little crickets were trying to tell me something. She believed that there was no better thermometer than a chirping cricket.

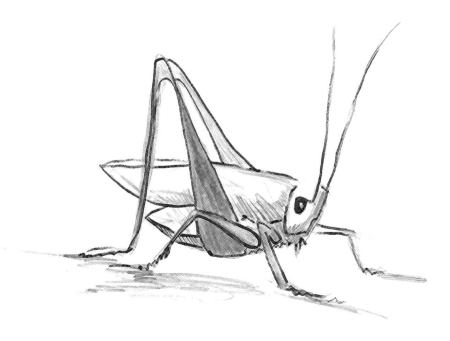

She was right. But here's something grandma didn't know: only male crickets chirp. They make those unique chirping sounds by rubbing the bottom of one wing on the top of the other in order to attract females during mating season. The warmer the air, the more frequent the chirps. In fact, chirping crickets don't chirp at all when the temperature is below twelve degrees Celsius.

To calculate the air temperature, you'll need a watch and you should make sure the house is very quiet, because you're going to be counting cricket chirps. Count the number of chirps in fourteen seconds; add forty and you have the air temperature in Fahrenheit. Of course, that's the temperature where the cricket is. The temperature may be a few degrees warmer where you are because crickets tend to stay low, often in the grass and in the shade. Now, when I was very young, the temperature was still measured in degrees Fahrenheit. To calculate the temperature in Celsius, count the number of chirps in twenty-five seconds, divide by three, and then add four. Give it a try; it's fun, and it'll put your math skills to the test.

Rain before seven, fine by eleven.

GRANDMA WAS AN early riser. In fact I don't think I ever heard an alarm clock go off in her bedroom. She was up with the roosters. When the day got off to a rainy start, she would tell us that before long, the rain would stop: "Rain before seven, fine by eleven." This is one of those weather sayings that are not always correct, but it is often right, and can be explained with a quick look at a weather map.

Some systems are more active at certain times of the day. A cold front is often made more powerful with the benefit of daytime heating, especially if the sun is out. Cold fronts trigger those dramatic late-afternoon thunderstorms that we sometimes experience at the end of a hot summer day. A few hours after sunset, a lot of that energy subsides.

A warm front on the other hand has a much more gentle approach. It's a slower moving wide band of moisture that comes in to replace cooler air. Because the warm air is less dense, it slides up and over the colder air; condensation occurs and rain falls behind the front. That process is helped along by the cooling of the air after sunset. The slow-moving system can take as long as twelve hours to move through, so if the rain began at sunset, it should be on its way out shortly after sunrise. Once the front passes, the sky clears and the air pressure rises. Temperatures also rise as warm air replaces cold air.

So if you wake up to the sound of light rain dancing on the roof, don't despair; the day might not be a writeoff. Grandma would tell you that there's a good a chance that rain could end before noon, with some sun to brighten the rest of your day.

Find out how far away the thunderstorm is by counting.

. . . .

Storms are natural phenomena that inspire strong reactions in both humans and animals. Some people love to watch them, standing at the window as the thunder and lightning crash all around. Some people choose to go outside, taking what many would consider an unhealthy risk in order to play in the rain. At the opposite end, some humans and animals have a fear of thunderstorms; astraphobia is a fear of thunder and lightning. I have a few mild phobias, but astraphobia is not one of them. I'm not sure where my love for a good thunder-and-lightning storm comes from; both my mother and Grandma were astraphobic. Each time a thunderstorm rolled in, they would reach for the holy water and sprinkle the windows with it. This was supposed to keep the lightning from striking the house. I guess it worked, but I digress.

Nighttime thunderstorms were my favourite. Once I'd heard the first rumbles of thunder, I would lie in bed very quietly, hoping the storm would get a little closer—close enough so that I could see the magnificent bolts of lightning light up my bedroom. Grandma knew that the greater the gap between the lightning and the thunder, the farther away the storm. If that gap narrowed, the storm was getting closer. One day, Mom told my sister and me that there was a way to figure out just how close.

Before I get to the formula, here's a brief science lesson. Sound travels through air at "the speed of sound," which is, officially, 331.3 metres per second in dry air at zero degrees Celsius. At a temperature like twenty-eight degrees Celsius, the speed is 346 metres per second. As you can see, the speed of sound changes depending on the temperature and the humidity, but if you want a round number, then 350 metres per second is a reasonable number to use. So sound travels one kilometre in roughly three seconds.

When you see a flash of lightning you can start counting seconds. When the thunder hits, stop counting and divide your number by three to see how far away the lightning struck. If it takes twelve seconds for the thunder to roll in, the storm is about four kilometres away. I don't know if this will take away any fear of thunder and lightning, but it might help pass the time while the storm rolls through.

If the dog is acting up,
there'll be thunder before long.

W E ALWAYS HAD at least two dogs on the family farm in Bainsville, Ontario. Over the years, we were blessed with a few excellent herding dogs as well as some that didn't have a clue. We loved them all. Aside from getting the cows into the barn for milking and back out after chores, the dogs were little weather prognosticators, especially in the summertime.

When summer's sultry heat started to build in the late afternoon, Grandma would turn to the dogs. She believed that they would let us know if a thunderstorm was approaching. Did they ever! We had one dog who would

start to pace and pant on a perfectly clear, sunny day an hour or more before we noticed darkening skies and realized that a storm was in the offing. This is pretty common; many dogs become agitated before a storm actually arrives. Storm-sensitive dogs realize early that a storm is brewing because they can smell it coming long before we know anything is going on.

It's been estimated that dogs, in general, have an olfactory sense ranging from one hundred thousand- to one million-times more sensitive than a human's. In some dog breeds, such as bloodhounds, the olfactory sense may be up to one hundred million-times greater than ours. Those adorable little noses are so sensitive that they can detect concentrations of chemicals in the low parts-per-million range. During a thunderstorm, lightning ionizes the air with the formation of ozone, which has a characteristic metallic scent. Experts believe that dogs detect this odour, while we remain oblivious to it. So Grandma was right. Man's best friend is also a pretty good weather forecaster.

If you have a rhododendron, you don't need a thermometer.

I F YOU WERE to randomly stop people on the street and ask them what their favourite season is, most would say summer. It's not to say that the fair season is without its hurdles, but they are a little less trying than winter's challenges. In nature, animals—and plants for that matter—face those challenges all the time. We could learn a thing or two by observing how they cope. Our ancestors watched carefully for signs of change as the cold set in and you can too. One of those signs might take place in your very own yard, if you're lucky enough to have a magnificent rhododendron.

Interestingly enough, rhododendrons have their own way of coping with the cold weather. These broad-leaved evergreens have always intrigued me with their temperature-sensitive leaf movements. Their thick, evergreen leaves begin to droop and curl as the temperature approaches freezing, and the colder the temperature, the tighter they curl and the more they droop.

Grandma wasn't the first to check the rhododendrons for an idea of how cold it was outside. Leaf movements in plants were first categorized by Charles Darwin in 1880. Darwin pointed out that many plant parts, and particularly leaves, move in response to a number of environmental factors, the most important of which are light intensity and direction, water content, and temperature. Years later, in 1933, a Japanese scientist studied the leaf-curling patterns of rhododendrons and made the important observation that their leaves could be kept from curling if he covered them with snow, thereby insulating them from the cold air. So it's contact with the cold air that makes the leaves curl under.

There are several different theories as to why this leaf-curling and drooping occurs in rhododendrons, but there is no debate that it is temperature-related. Many researchers feel that it's nature's way of protecting the leaves from drying up due to freezing temperatures and cold winter winds. Others believe that the curling protects delicate cell membranes from damage caused by rapid thawing following a freeze. Their theory is that the curled leaves hanging vertically thaw much slower than flat leaves held horizontally, and this protects the cells from freeze damage. Regardless of the reason, I'm happy just to know that they curl when it's cold and leave it at that. But some people take it a step further. They claim that they can tell what the temperature is by the size of the curl! Imagine: a rhododendron thermometer.

Morning dew on the grass, rain will never come to pass.

I KNOW THERE'S a time and a place for footwear, but I've always loved to go barefoot. There's something very liberating about not wearing shoes. It used to drive Mom crazy. She was always worried that I'd step on something sharp and cut my foot. She had a point. It is wiser to wear shoes, but not nearly as much fun. I loved the various textures I'd come across between the back door and the garden. First thing in the morning, the old cement landing usually felt quite cool underfoot. My next few steps took me across some gravel that, to be

⌒ CINDY DAY ⌒

honest, did pinch a little. Then on to the grass; I loved how it tickled my toes. I remember that, in the summer, I would often come back into the house with wet feet. When I tried to explain that I had stayed on the lawn but the grass was wet, Grandma would smile and say it was going to be a good day to work outside: "Morning dew on the grass, rain will never come to pass." I couldn't imagine where the wetness had come from on such a clear morning.

Dew forms when the air temperature falls to the "dew point temperature." An air mass has two temperatures: the actual air temperature and the infamous dew point temperature. "Dew point" is defined as the point to which the air must be cooled in order for the air mass to become saturated with moisture. Once the air is saturated, condensation occurs on objects like the hood of a car or blades of grass. The necessary drop in temperature is most likely to take place on a clear, windless night. Those conditions exist in the centre of a fair weather system. When a fair weather system is overhead, you can be pretty sure that it will take another day or so before it pulls through, allowing the next weather system to move in. This almost guarantees a dry day.

When seaweed becomes a little damp, there's rain on the way.

· ·

HAVING GROWN UP on a dairy farm, I'm partial to the pastoral beauty of our sprawling rural areas: acres of lush fields laid out like squares on a quilt; rolling hills with streams and rivers meandering until they seem to drop off the earth.

After I left the farm, I discovered a new love: our magnificent coastline. I remember the first time I came across a small fishing village perched along the rugged shore. The homes were modest but well kept. Behind the houses were small sheds, and each one had a line stretched from it to a stately tree a few dozen metres away. Some of the lines were empty, but others were draped with something that looked like darkly coloured rags or thick rope. I was far too curious to drive away without finding out what was going on. (I get that curiosity gene from Grandma.)

As I made my way down the gravel road that led to the village, a very pleasant older lady greeted me. She was happy to explain that she and many others were drying seaweed. After some pretty steady rain, the weather had cleared a couple days earlier and conditions were good for drying. When I asked her when she would take it inside, she said she would watch the seaweed. Fishing families know that when the seaweed becomes a little damp, there's rain on the way.

I had never heard that one before, but it made perfect sense; it has everything to do with the salt remaining on the surface of the seaweed. Salt is hygroscopic, which means it will absorb moisture when the air is humid. The air's relative humidity rises before a rain system moves in.

(You might find this interesting. The inventor of the first hygrometer was Nicholas of Cusa, a German cardinal and philosopher who, in 1450, experimented with placing some sheep's wool on a set of scales and then monitoring the change in the wool's weight as it absorbed moisture from the air, or lost moisture on a dry day. He rightly concluded that the amount of water vapour in the air changed, and that the change could be measured.)

When ladybugs swarm, expect a day that's warm.

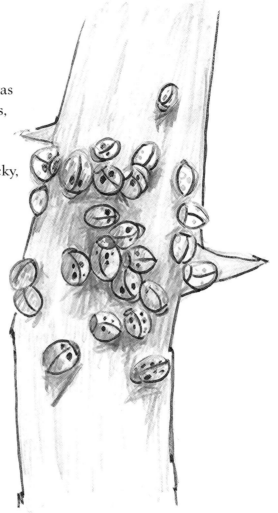

AS A KID growing up on the farm, I was always fascinated by bugs and insects, and there were always lots around. My favourite insect was the lovely ladybug. Many cultures consider ladybugs to be lucky, and have nursery rhymes or local names for the insect that reflect this. In many countries, including Russia, Turkey, and Italy, the sight of a ladybug is either a call to make a wish or a sign that a wish will soon be granted. I'm not sure if Grandma considered ladybugs to be lucky, but she did consider them to be useful; many species feed on aphids or scale insects, which are pests in gardens, agricultural fields, and orchards. More importantly, Grandma turned to the ladybug during the summer to gauge the temperature for the day. She used to say, "When ladybugs swarm, expect a day that's warm."

It turns out that little rhyme is quite accurate. Ladybugs are cold-blooded, meaning weather conditions affect their internal temperature, behaviour, and lifestyle. Ladybugs are only active within a particular temperature range. Generally, ladybugs won't even fly if the temperature is below thirteen degrees Celsius. At the other end of the scale, they become very active when it gets too hot. Ladybugs store heat in their shells. If it gets too warm, they start flying to dissipate the heat. Of course, this means it is already warm.

I recently came across an interesting little story about a Welsh custom that dates back to the late 1800s, which involved using ladybugs to predict the weather. If you were able to catch one and hold it in your hand, you would chant a poem, then throw the insect up in the air. If the ladybug fell to the ground, then it would rain; but if it flew away, the outlook was sunny.

When the ditch and pond offend the nose, then look for rain and stormy blows.

WHEN I THINK back to those magical years I spent growing up on the farm, I can't help but remember the smells. Like the smell when I walked into the milk house, or the smell of fresh hay. Of course other smells were, shall we say, less pleasant.

Nature itself has a scent; a combination of oils released from plants, gases from swamps, and of course the more pleasant, sweet aroma of wildflowers. You've probably noticed that some days, those smells are a little more pungent. When Grandma found that the garden flowers smelled sweeter, or that the smell coming from the nearby creek behind the house was a little sharper than usual, she would point out that there was a storm coming: "When the ditch and pond offend the nose, then look for rain and stormy blows."

There's a good reason for that. Smells travel more easily when there is more moisture in the air. The amount of moisture rises ahead of a weather maker. The next time you're out for a walk, take a deep breath and concentrate on what you smell. If you smell moist, earthy odors, there could be rain on the way. You see, plants release their waste in a low pressure atmosphere, generating gases that smell a little bit like compost. That smell could also be attributed to a nearby swamp. Swamps will release gasses just before a storm because of the lower pressure, which leads to unpleasant smells. If you think the flowers are especially fragrant, turn around and go back for your umbrella: flowers smell best just before a rain because scents are stronger in moist air. So if someone tells you they can smell the rain coming, you can tell them why.

Nobody can water the garden like Mother Nature can.

FOR MY SISTER Monique and me, the countdown to summer vacation started just after the Easter weekend. While many of the kids in class were planning on going to day camps or travelling to a big city to visit relatives, we knew that our summer would be spent in the field. I guess that's why we really didn't mind if it rained once in a while; it gave us a break. Our friends from the city on the other hand couldn't have cared less if we didn't get a drop of rain all summer. They were on vacation. I guess some people don't stop to think about where their food comes from and the important role the weather (and yes that includes rain) plays. On the farm, we didn't hope for sunny weekends and rain during the week; we prayed for just enough rain after the crops went in and not too much when it was time to harvest. Our livelihoods depended on it. On a smaller scale, so did our dinner.

We always had a fabulous vegetable garden. Mom and Grandma had very green thumbs, but they would often defer to Mother Nature when it came to taking credit for the garden's bounty. Grandma used to say that you could water the garden all you wanted, but it wasn't going to be nearly as good as the water that came from the clouds. It turns out she was right. Like snow, rain contains nitrogen compounds. It is estimated that two to ten kilograms of nitrogen are deposited per hectare of land as a result of snow and rain. As I mentioned earlier, most of this nitrogen comes from emissions from industrial manufacturing and the burning of fossil fuels; the rest comes from lightning fixing atmospheric nitrogen, which makes up seventy per cent of air.

In everyone's life, a little rain must fall…thankfully!

Cobwebs on the lawn.

HAVE YOU EVER stopped to admire a spider's handiwork? Every once in a while Grandma would come in from her walk to the hen house to collect eggs and declare, "It's going to be a hot, sunny day—the spiders were busy last night." She was usually right. An early morning display of intricate cobwebs across the lawn does indicate a sunny and unusually hot summer day. But how could the spiders know?

Well, they don't really know what kind of weather is coming, but they do know what weather conditions they need to build their webs. Spiders are not able to spin their silk when it's windy. On a clear, windless night, they get to work. In the summer, clear nights with little or no wind occur when there's a

high pressure system directly overhead. The following day, as the high drifts eastward (remember that weather almost always moves west to east), the wind comes around to the south, and combined with the summer sun, serves up a scorcher of a day. Clear, windless nights often allow nighttime temperatures to cool rather quickly. By sunrise, dew will have developed. That's when the cobwebs really stand out. The silks trap the beads of dew and glisten in the morning sun. So the next time you're out for a stroll, look around, you never know what little mysteries Mother Nature has laid out for you to discover.

What St. Swithin says about summer.

THERE ARE A number of calendar days that Grandma never let got by quietly. One of them was July 15, the feast day of St. Swithin. Perhaps you've heard of him?

"St. Swithin's Day, if ye do rain, for forty days it will remain. St. Swithin's Day, an' ye be fair, for forty days 'twill rain nae mair."

I don't think anyone talks like that anymore, but Grandma certainly paid attention to the message. The last thing a farmer wanted in mid-July was rain, especially if that rain was going to last for forty days.

So who was this man? St. Swithin, or more properly, Swithun, was a Saxon bishop of Winchester. He was famous for giving charitable gifts and building churches. The legend says that as the bishop lay on his deathbed, he asked to be buried outdoors, where he would be trodden on and rained on. He didn't want any special treatment. For nine years, his wishes were followed, but then, the monks of Winchester attempted to move his remains to a splendid shrine inside the cathedral on, you guessed it, July 15, back in the year 971. Legend has it that a heavy rain storm rolled across the region during the ceremony and that it continued raining for forty days.

This led to the old wives' tale that if it rains on St. Swithin's Day, it will rain for the next forty days, but a fine fifteenth of July will be followed by forty days of fine weather.

However, according to the data compiled by a researcher at the meteorological office in London, England, this old wives' tale is nothing other than a myth. It has been put to the test on fifty-five occasions when it was wet on St. Swithin's Day and forty days of rain did not follow. Well, I think I speak for the farmers out there when I say that's a good thing.

If there are raindrops on the clothesline in the morning, there will be rain before the end of the day.

GRANDMA WAS A master weather observer, and over time that led to her very high accuracy rate when it came to short-term forecasting. She wasn't alone. Decades ago, many people honed these skills out of necessity. Farming is as much about the weather as it is about the hard work. Now, sometimes Grandma's forecasts had nothing to do with planning field work. Take laundry day for example. Long before electric dryers, the clothes had to be hung outside to dry. Farming is dirty work so there was always lots of clothes to

wash and dry. The weather had to be just right. While the clothesline held the jeans and work shirts, it also held the key to the weather for the day ahead.

Grandma always said, "If there are raindrops on the clothesline in the morning, there will be rain before the end of the day." I've watched this one over the years, and it's proven to be fairly accurate. It's one of the many weather sayings that reflect the increased humidity in the air that often precedes rainy weather. When the humidity is high, water condenses onto

colder surfaces, like the clothesline, that have cooled overnight. When the moist air touches the clothesline, it becomes cooled below its dew point and condensation occurs. The higher humidity is usually a sign that rain is imminent.

A sick dog might be a sign of rain.

G RANDMA WAS VERY aware of how animals reacted to weather changes, including the family pets. I paid close attention to her observations, and I am pleased to report that now I have a "weather beagle." Her name is Bix and she has not yet led me astray. She doesn't waste her time with your run-of-the-mill storms, but she's right on top of the big ones.

The first time I picked up on her weather sensitivities was the day before Hurricane Juan rolled into town, on September 28, 2003. She didn't seem to be herself. She was constantly at the door, wanting out. Each time she went out, she headed for the most lush area of the lawn and chomped away on the grass. By evening, about six hours before Juan arrived, her stomach turned. After that, I watched carefully each time she indulged in "salad." Without fail, her stomach turned and a storm rolled through six to eight hours later.

At first I thought it might be coincidence, but after monitoring her "episodes" I realized there was something to it. It's not been scientifically proven, but many vets believe that animals are quite sensitive to sudden changes in air pressure, and that a rapid drop in atmospheric pressure can cause some discomfort in animals, including dogs. As a storm approaches and the pressure falls, your pet might experience a nauseous feeling. A dog's stomach has neuro-receptors that respond to what it has eaten. Grass is believed to be emetic; it induces vomiting. So the next time Fido spends time chomping on grass, check the barometric pressure…you too might have a weather pooch on your hands.

The turkey's breastbone gives hints for the winter.

. . . .

Thanksgiving is one of my favourite holidays. What's not to like about it? The food is fabulous, the countryside is breathtaking in its stunning fall cloak, family and friends come together and…we get to find out what the upcoming winter will be like. Well, at least that was always the case at our house on Thanksgiving. Mom carved the turkey, but Grandma presented the weather forecast at the end of the meal—with the help of the bird.

After enough meat was removed from the turkey to get to the breastbone, Grandma prepared her prognostication. According to her, the length of the breastbone indicates the length of the coming winter while the colour of the breastbone indicates its severity. A plain white breastbone means the winter will be mild, and a darker, mottled breastbone indicates a more severe winter— the more mottled the breastbone the more severe the winter will be. And there's more: purple tips are a sure sign of a cold spring.

Not long ago, I met a farmer who claimed he could explain why. Apparently the darker colour meant that the bird had absorbed a lot of oil, which acts as a natural protection against the cold. The darker the colouring, the tougher the winter ahead would be.

So does this really work? Finding out could make your Thanksgiving dinner lots of fun. Choose someone to be the breastbone examiner; have someone document the findings, keep the breastbone for proof, and check back in the spring.

When flowers continue to bloom well into the fall, we're in for a snowy winter.

. . . .

IN METEOROLOGICAL CIRCLES autumn is known as a shoulder season. It's like a buffer between summer and winter. In some parts of the country it can be rather brief. One or two good blasts of arctic air and voila, winter settles in. In the Maritimes, fall can be as nice as summer and often is. We're surrounded by very large bodies of water that take a long time to cool down, so we're often treated to lovely fall days. Some years, this mild weather extends well into November. That might be nice for humans, but what about nature's natural cycles? Some years I recall seeing wild strawberries in bloom along the side of the road on Halloween, and I can remember seeing Easter lilies in bloom in the garden on Remembrance Day. Grandma would say "when flowers continue to bloom well into the fall, we're in for a snowy winter."

Scientifically, a very warm November that tricks nature into thinking it's spring is simply an indication that the jet stream has not yet shifted and that we're still on the warmer side of that influential river of air in the upper atmosphere. A late shift could result in proximity problems for us; the final resting position could be too close for comfort. The jet stream could easily slip just far enough south to give us colder weather, but stay close enough to bring the storms to our region, which means we could be doing lots of shovelling. So while May flowers can bring cheer to an otherwise dull November day, they might also bring jeer before long.

Acorns on the ground before Michaelmas, snow before Christmas.

S EPTEMBER CAN BE a tough month: summer vacations are over, the kids are back in school, and the garden is starting to look a little tired. Despite all of that, Grandma found something to celebrate. On the twenty-ninth, she'd be the first to wish everyone a Happy Michaelmas. You see, September 29 is the feast day of Saint Michael, the archangel, patron saint of the sea and maritime lands, and of ships and boatmen.

Traditionally, Michaelmas was the last day of the harvest season. In many areas of the country today, harvest is still in full swing at the end of September. Our crops have been modified for higher yields and in some cases require a longer growing season. And then there's the issue of our changing weather.

As you know, Grandma watched the calendar quite closely, and on September 29 we headed out to assess the acorn situation. According to weather lore—and Grandma of course—if there were lots acorns on the ground on Michaelmas, there would be snow on the ground before Christmas.

While we were chasing acorns, some people were eating goose. In many parts of Europe, September 29 was called Goose Day: "Eat goose on Michaelmas Day, want not for money all year." Doing an acorn count was fun, but had Grandma told us about Goose Day, things could have been very different.

The more berries on the mountain ash tree, the more severe the winter will be.

· ·

FOR SOME, AUTUMN is a welcome break from summer's high heat and humidity. Those cooler temperatures eventually lead to some pretty incredible fall colours. By the time Thanksgiving rolls around, the palette usually includes vibrant reds, shocking oranges, and standout golds. Grandma appreciated a spectacular display of fall foliage as much as the next person, but the colour she was looking for was dangling from the branches of the mountain

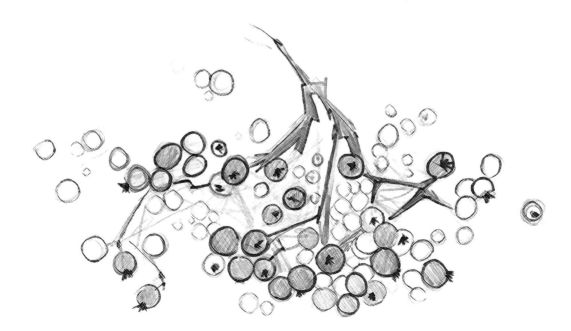

ash tree. She, like so many others, would seek out the mountain ash in the fall for a look ahead to the coming winter. She believed that a heavy crop of berries pointed to a winter with lots of snow and high winds: "The more berries on a mountain ash tree, the more severe the winter will be."

In some parts of the world, the mountain ash is known as the rowan tree. Despite the different name, it remains very connected to the weather. In Finland if there were a large amount of rowan flowers in the spring, the rye harvest would be plentiful in the fall. If the tree flowered twice a year there would be lots of potatoes and, oddly enough, many weddings that autumn. In some lands, winter would begin when the birds had eaten the last of the berries on the mountain ash or rowan tree. In Sweden, if the rowan grew pale and lost its colour, the fall and winter would bring illness and suffering.

The mountain ash is not just a showy weather predictor but is quite a sacred tree in many regions of the world. Branches from a rowan tree were blessed at the local church then carried onto ships to avoid storms. Some people kept branches in their homes to protect them from lightning.

While I've not been able to find a correlation between an abundance of fall berries and a harsh winter, the weather folklore surrounding this fruitful tree is undeniable, and Grandma would remind me not to question hundreds of years of careful observation.

Woolly caterpillar

· · · · · · · · · · · · · ·

J UST WHEN YOU thought the groundhog had dibs on weather folklore, along comes the woolly bear caterpillar. In long-term forecasting, one of the most talked about insects is the woolly caterpillar. Each year, as summer winds down, people start noticing them and paying close attention to their stripes. Grandma was always keen to spot a caterpillar slithering across the front step. She, like many others, believed that the wider the middle reddish-brown section, the milder the winter would be. It has become one of the most popular pieces of fall folklore around. A lot of pressure for a little caterpillar.

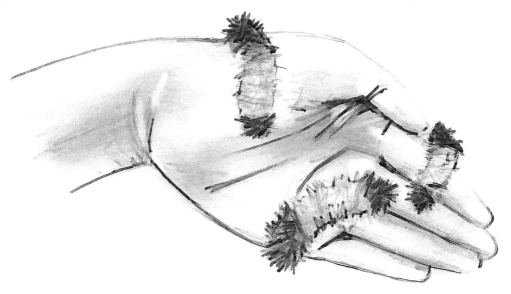

I wish I had a solid scientific explanation for this one, but the truth behind the woolly caterpillar's band width actually has more to do with its age than predicting the weather. As it prepares for the winter, the caterpillar molts, becoming less black and more reddish-brown as it ages.

That doesn't stop the good people of Ohio from celebrating the little fella. The mythical qualities attributed to the caterpillar, commonly know as the "woolly bear," has led to an annual Woolly Bear Festival in Vermilion, Ohio, on Lake Erie. Every fall, people young and old, many dressed in woolly bear costumes, gather for a fun-filled day. Why not? The groundhog has his day!

When the leaves are backwards on the trees, there's rain on the way.

· ·

W HILE DOCTOR DOOLITTLE talked to the animals, my rural upbringing taught me to observe the animals. Birds, cats, dogs, and cows can all teach us something about the incoming weather. But nature, too, is worth a look. Grandma always had her eye on the trees. She could tell if she should hang laundry on the line to dry by the way the leaves looked. She would say, "When the leaves are backwards on the trees, there's rain on the way."

As a child, I found the notion of the leaves being backwards on the trees quite perplexing. I came to learn that what Grandma meant to say was if you could see a leaf's underside, there was rain coming. Still quite baffling.

Not surprising was the fact that this too can be explained scientifically. Atmospheric conditions are responsible, mainly wind. Across North America, our prevailing winds are westerly. In the winter, they become a little more northwesterly and in the summer, a little more southwesterly. That being said, as the leaves grow on the trees they are held down by that prevailing westerly wind. When a rainstorm approaches, counter-clockwise winds around its centre produce easterly winds. That wind goes against the flow. So the east wind will pick up the leaf and flip it over. As the sky darkens, the undersides of leaves become more visible and the mosaic patterns of lighter green mixed with the darker green become more enhanced. By observing these leaf patterns along with a darkening sky and wind, you can predict rain is on the way in the immediate future.

Clear moon, frost soon.

IS THERE ANYTHING more mysterious than a large, pendulous full moon casting its golden light over the countryside? A full moon is magnificent regardless of where you live, but in rural areas, where there is little other light to compete with, it's magical. I can remember coming home as a very young girl from my uncle Keith's house one Sunday evening when Dad decided to show us just how bright the moonlight was. We had just turned on to the Westley side road when Dad turned off the car lights; back then you could do that. Mom was not amused, but my sister and I thought it was the coolest thing. Dad didn't drive very far with the lights off, but I remember being amazed at how clearly we could see the road and the fields around us. When we got home, Dad told us that many years ago, before tractors had headlights, farmers used to take advantage of the light of the full moon to finish up their work in the fields.

Grandma wasn't as concerned with the light of the full moon as she was with the timing of it in the fall. Some years, the garden tomatoes were late ripening and not quite ready to be harvested by the time frost was in the air. Grandma determined the chance of frost by the moon, and would cover the plants with wet newspaper or an old sheet to protect them if needed. In the fall, when the full moon shone brightly, she would remind us that a clear moon meant frost soon. She was right. If you're able to see the full moon, it has to be clear above. Clear skies allow the earth's heat to radiate into the upper atmosphere. With no clouds to hold that heat in, temperatures fall quickly. In the fall, that drop in temperature often results in frost. So while it is true that a change in the moon can't change the weather, a good look at the moon can help forecast it.

If the hornets build their nests high off the ground, it's going to be a snowy winter.

A FALL STROLL after most of the leaves were off the trees could give Grandma some insight into how much snow we could expect through the coming winter. Every year, sometime between Labour Day and Thanksgiving, Grandma would walk in the house with a big smile on her face and a winter snowfall prediction. She had obviously just come across a hornets' nest. Many old-timers believe that if the hornets build their nests high off the ground, it's going to be a snowy winter, but if those nests are close to or on the ground there will be little or no snow. Because so many people believe this to be true, I did all that I could to find some scientific explanation, but alas! No such luck. I can, however, explain why some people find nests high in the trees while others find them on the ground in the same area during the same season.

There are many different types of hornets. For example, the term "yellow jacket" refers to a number of different species of wasp. These wasps are ground-nesting wasps, regardless of the weather. You'll find their nests in rotted tree stumps, rodent burrows, or even holes in the foundation of an old shed. Then there is a group of aerial-nesting yellow jackets. They build their paper nests on the eaves of a building or hanging from the limb of a tree.

While I couldn't find a link between the location of a wasp's nest and the weather for the coming winter, I did find a weather connection. When it comes to building the nest, current weather is a big factor in the wasps' choice of location. Warm, dry weather is ideal so if you live in a rural area during summer, don't be surprised to see a wasp colony under your front porch.

Onion skins very thin, mild winter coming in; onion skins thick and tough, coming winter will be cold and rough.

. . . .

MY PARENTS AND grandparents worked from sun-up to sundown and then some. A lot of that work was on the land. Farmers work the land to feed the animals, but life on the farm often includes a huge vegetable garden. Grandma was very proud of her garden. I have such fond memories of her handing me the "big yellow garden bowl" and a knife and sending me off to pick beans and cut some leaf lettuce for supper. Things got interesting late in the season when it was time to harvest the onions. Grandma always looked forward to slicing through the onion to check out the thickness of its skin. According to folklore, "Onion skins very thin, mild winter coming in; onion skins thick and tough, coming winter will be cold and rough."

The belief that onions can help predict the weather comes from the Gauls, who observed that many layers of thick skin signaled the coming of a harsh winter. Some would argue that it's nature's way of preparing for the winter ahead. Scientifically, this one is difficult to explain. But who are we to question an observation that goes back to 300 BC?

I often wondered if this saying was Grandma's way of getting us excited about eating onions. The next time you head off to your local farmers' market with the kids, pick up some onions. When you get home, slice into them and note the thickness of the skins. The kids can write down the findings on a calendar and you can verify the results in the spring. You have nothing to lose and you might end up with a nice big pot of French onion soup.

An open pine cone is a sign of fair weather.

. . . .

OVER TIME, I'M sure you've come to realize that some of Mother Nature's signs are more reliable that others. One of the most reliable of all natural weather indicators are pine cones. For decades they've been used to forecast the weather; mind you, it's a short-term forecast, but I digress. I used to do quite a bit of camping and I always made a point of collecting a few pine cones on the way to the campsite. I'd set them up on a flat surface around the camp area. After a while I'd check in. An open pinecone indicated that the air was really dry and that the weather should be good for awhile. However, if the cone stayed closed it meant that the humidity was high and stormy weather might be on the horizon. Not all species of pine cones do this. There is, however, an advantage for those that do. Pine cone seeds are spread by the wind and on a rainy day they may not go as far. When the air is dryer, the outside of the scale shrinks more than the inside and the scales open up allowing the seeds to get out. Mother Nature is pretty crafty!

The cones don't have to be on the tree to do this—that really annoyed Grandma. You see, every fall we would set out to pick pine cones so Grandma could make a wreath. To make sure the cones would stay open, and not close up the next time a storm approached, she'd pop them in the oven for awhile. By roasting in a slow oven for a couple of hours they would become completely dry. With all the water removed, they would stay open and pretty for Grandma's wreath.

If the squirrel's tail is bushy,
it's going to be a cold winter.

· ·

ONE DAY I WATCHED as a little brown squirrel tried to get at some seed in a bird feeder that Mom had hung from a tree branch. The feeder was shaped like a little house and had quite a steep roof. Well, that little fella was not going to let the roof keep him from his meal. He made his way up the tree then onto the branch. He then jumped from that branch onto the top of the feeder and slowly inched his way to the edge of the roof, and just as he was about to leap onto the bed of seed below—down he went! The feeder had tipped just enough. That didn't dissuade him one bit. Back up the tree he went. He repeated the process a half-dozen times before he managed to twist his little body and land on the seed as the feeder started to spill over. I think squirrels can teach people that there is no obstacle that can't be surmounted.

Grandma loved to watch the squirrels on the farm; they were quite entertaining, swiftly making their way from tree to tree. As winter approached, the squirrels became a valuable forecasting tool. Grandma believed that if the squirrel was nice and plump and had a big bushy tail, we were in for a hard, cold winter. So is this a sure way to predict the weather, or just a case of an over-indulgent squirrel? I am afraid that the scientific community is not on board with this one. The idea that you can predict a harsh, cold winter by the size of a squirrel's tail is mostly folklore. But Grandma wasn't the only one to turn to the squirrel for a late-fall forecast. The Pennsylvania Dutch also believe that when the squirrel's tail is extra bushy and its fur is nice and thick, the winter weather will be severe. I think it's something worth watching, even if only for the sheer entertainment value.

When swallows fly high, the weather will be dry; when birds fly low, expect rain and a blow.

SCIENTISTS DON'T REALLY know how, but it's become very obvious that birds can sense changing barometric pressure. One theory is that birds may be able to detect that drop in pressure through their inner ear. It makes sense really. We're able to detect large changes in air pressure in our own inner ear when we make a fast change in altitude; that's when our ears "pop."

Aside from sensing the drop in air pressure, low pressure will impact a bird's routine. Lower air pressure means that the air is less dense, therefore harder to fly in. That's one reason we see birds gathering close to the ground before a

storm. But there's more. As a weather system approaches, the upper winds are the first to increase. Strong upper winds make it more difficult for the birds to fly at that level so they stay low, closer to the ground where the wind is not yet as strong. Conversely, the wind is usually quite light when a fair weather system is overhead, and the pressure is high too. It is much easier for the birds to manoeuvre high in the sky when the wind is light and the air is dry. As is the case with humans, birds will take the path of least resistance.

Will it rain? Check the bubbles in your morning tea.

· · · · · · · · · · · · · · · · · ·

THERE'S SOMETHING SOOTHING about a nice, hot cup of tea. When things got a little hectic at the farm, I remember hearing Mom or Grandma say, "Let's sit down and have a cup of tea." It seemed to bring things into focus. Tea was also the first thing offered to someone who came to visit, and the last thing consumed at the table after a lovely dinner. Some people drink it black, others "take it" with a little milk and sugar. Grandma liked a splash of milk in her tea, but she didn't dare add it until she had checked the bubbles.

Grandma could tell you what the weather was going to be by how the bubbles were displayed in the cup. She believed that if the bubbles gathered

in the middle of the cup, there was sunny, dry weather on the horizon. If the bubbles clung to the sides of the cup, it would be cloudy and damp.

High pressure systems serve up fair weather. During a period of high pressure, air is pushing down on the earth. The extra weight exerted by the air can make the surface of the tea slightly concave so that the bubbles on the surface will slide into the middle of the cup. On the other hand, low air pressure during foul weather will make the surface of the tea slightly convex and the bubbles will slide down and stick to the edges of the cup, forming a ring. So maybe all those people you see staring into their cups are aspiring meteorologists.

Aches and pains, coming rains.

· ·

EVERY NOW AND again, someone asks me how I can possibly remember all of Grandma's little gems. Well, I heard them over and over, year after year, and not just in passing. Grandma always made a big deal about weather lore. Sometimes she would tell a story about how her father or even grandfather would watch for the very same signs.

Having said that, here's one weather saying that I didn't hear until Grandma got a little bit older: "Aches and pains, coming rains." I'm starting to understand why; if you have even the slightest bit of arthritis, you probably do too. Grandma had noticed that before a weather system arrived, her joints ached a little. Over the years, a number of medical studies have been able to find a link between the barometric pressure and those achy joints.

Air pressure falls ahead of a storm. Air pressure can be defined as the weight of the air exerted on the earth and of course on all of us. A drop in this weight or pressure allows blood vessels in our body to dilate or expand slightly. If your joints are already a little swollen with arthritis, the extra expansion makes things pretty tight in those small spaces. This also has the effect of aggravating already-irritated nerves near corns or even cavities. Who knew that your corns could give you a heads-up on the day's weather?

The higher the clouds, the better the weather.

F EW THINGS ARE as much fun to watch as clouds. Some of my earliest childhood memories involve lying on the grass in front of the farmhouse with Mom and my sister Monique. We could spend hours looking for shapes in the passing clouds. It was such a fun game. Some days there seemed to be all kinds of dogs drifting by; other days we were sure there were angels with huge billowy wings looking down on us. Every now and again we saw faces in the clouds. There are all types of clouds out there, and while they can be quite intriguing all year long, kids seem love the summertime puffiness of the cumulus clouds.

Grandma didn't play the "cloud game" with us, but she didn't ignore the clouds either. She could tell what kind of weather was coming by the height of the clouds. She would say, "The higher the cloud, the better the weather." She was right. Clouds have always played a big part in short-term weather forecasting. When the air is dry, the condensation level is higher off the ground than it would be if the air was muggy, so the base or bottom of the clouds forms higher in the sky. Dry air comes with a fair weather system. The opposite is also true. When the air is humid, the condensation level is low and clouds form closer to the ground. Humid air moves in ahead of rain or snow. So before you head out in the canoe or take the kids for a hike, a quick cloud check might not be a bad idea.

Mackerel sky, mackerel sky, never long wet, never long dry.

. . . .

D ID YOU EVER look up at the sky only to see it completely covered by what looked like fish scales? Or maybe you're not into fish and those lumps of clouds reminded you of buttermilk. According to weather lore, that very distinctive cloud cover— known in some circles as a mackerel sky and in others as a buttermilk sky—could help you plan your day. The phrase "mackerel sky" comes from the idea that the lumps of clouds are similar to the markings or scales of an adult mackerel fish. The reference to buttermilk is self-explanatory and in my opinion not nearly as tasty. Grandma called it a mackerel sky, and she often repeated this little saying: "Mackerel sky, mackerel sky, never long wet, never long dry."

A mackerel or buttermilk sky is an indicator of moisture and instability at intermediate levels, generally between 2,400 and 6,000 metres. If the lower atmosphere is stable and no moist air moves in, the weather will most likely remain dry. Should a few showers develop, they won't amount to much; you would need some moisture in the lower levels of the atmosphere for a more widespread rain to occur. Mackerel sky, therefore, is usually only used in reference to cirrocumulus or altocumulus clouds when they cover the majority of or the entire sky. Either way, it's never long wet and never long dry.

If the moon looks sick,
it's going to rain.

. . . .

WHEN PEOPLE ARE a little under the weather (pardon the pun), you can usually tell they're not well just by looking at them. Grandma would tell you that she knew there was rain coming when the moon looked sick. Imagine! A sick moon.

On a perfectly clear night, the moon can be spectacular. Sometimes it's so clear that you can make out the dried ocean beds on the moon's surface; other nights it's as though you're looking at the moon through a piece of waxed paper. At least that's how Grandma described it. You know the look; it's almost blurred or a little fuzzy. That's the sick moon. The moon looks like that when thickening cirrus clouds have moved in; you're looking at the moon through the ice crystals that make up those clouds. That cirrus cloud is a forerunner of wet weather; it can be rain or snow. Grandma had figured out that rainfall would usually come about twenty-four hours after she had spotted the sick moon. Over the years, I've tested this one and it's rarely been wrong. That gives new meaning to being "under the weather."

When it comes to the wind, east is least.

· ·

WIND. BY DEFINITION, it's simply air in motion. The wind brings us different sorts of weather. It also affects our lives when it blows strongly, and even when it doesn't. We curse it on a cold winter's day and we long for it during the hot, sultry days of summer. You might think that wind is just wind, but not so. In many parts of the world, people name their winds. Right here in Canada we have the famous Chinook winds that blow in during the winter and treat residents of southern Alberta to a welcome reprieve from the cold. Some of the wildest winds in Canada blow over the Cape Breton Highlands. There, strong southeasterlies can reach hurricane force as they funnel between the mountains and intensify before gushing out on the other side. These winds are known as *"les suetes"* winds and can peak at over two hundred kilometres per hour! There's also *"la Bise."* Grandma often talked of this "Bise," a cold dry wind that blows from the north in the mountainous regions of southeastern France and western Switzerland during the winter months. The Bise is always accompanied by heavy cloud. There's also the Maestro: the name given to northwesterly winds over the Adriatic Sea, the Ionian Sea, and coastal regions of Sardinia and Corsica. And we can't forget the Santa Anna winds: hot, dry, blustery winds that blow from the east or northeast over southern California. They are most common in winter and have been known to kick up very impressive dust storms. Those are just a few examples of winds around the world.

The fact that people have given local winds a name speaks to the significance of wind. Grandma was always watching the wind. We often got updates at the dinner table. When the wind turned to the east, we got more than an update. Grandma always said that "east is least." She believed that nothing good came of an east wind. In the winter it brought freezing rain. In the summer, it pulled in fog, rain, and strong damaging gusts. Meteorologically speaking, an east wind is definitely a sign of unsettled weather. Let me set the stage by stating these facts: A low pressure system brings wet or stormy weather. The wind pattern around a low is always counter-clockwise in the northern hemisphere. In North America, most of our weather systems travel from west to east. Okay, now imagine that low, pushing toward your area. The counter-clockwise wind reaching out ahead of it is from the east. The so-called bad weather is still to come. That's why a developing east wind is a pretty sure sign that unsettled weather is on the way. A few decades ago, Bob Dylan coined the phrase: "the answer is blowing in the wind." Who knew that Grandma and Bob Dylan could have something in common!

Le trois fait le mois.

· · · · · · · · · · · · · · · ·

WHEN I SAT down to write this book, I knew there would be one or two of Grandma's favourite weather sayings that I would struggle with. This is one of them: *Le trois fait le mois.* Many of the French language proverbs that are so popular in Quebec and among the Acadian population across the Maritimes came from France. They've been around for centuries and somehow continue to thrive in today's fast-paced social networking world. I only speak from personal experience when I say that I believe the French people continue to embrace their rich storytelling past. I often meet children who are very well versed on *les dictons*, or weather proverbs. I think it's fabulous and I hope it will continue.

Roughly translated, *le trois fait le mois* means the weather on the third day of the month will prevail until you turn the calendar page. For example, if you wake up to wind and rain on the third, and it's quite relentless all day, you should expect, at least according to Grandma, a windy wet month. But there's more, and it gives Mother Nature a chance to redeem herself: "*Le cinq le defait*"—the weather on the fifth can override the weather on the third. So, if the sun comes out on the fifth, the month will in fact be quite nice and not rainy as previously expected.

This may very well be the work of an eternal optimist, but it can be explained. Every area of low pressure is followed by a fair weather system, so chances are that if it was raining on the third, that system will have moved out by the fifth. That's as close as I can get to connecting the science of meteorology to this little weather gem. Having said that, a few years ago I decided to put it to the test. I monitored the weather on the third day of each month for a year and compared it with overall conditions for that month. *Le trois fait le mois* was accurate ten of the twelve months.

When kitty washes behind her ears, we'll soon be tasting heaven's tears.

GRANDMA WAS MORE of a dog person, but she appreciated the fact that cats were invaluable on the farm; they kept the mice population at bay and they were good little weather forecasters. The connection between cats and the weather goes back centuries and has a very close to tie to the fishing industry. Cats were believed to have miraculous powers that could protect ships from dangerous weather. French fishermen in particular would watch their cats closely to predict weather changes. They believed that if the cat passed her paw behind her ears while grooming, it was going to rain. From that belief came this little rhyme: "When kitty washes behind her ears, we'll soon be tasting heaven's tears."

This can be explained with a quick weather lesson. Before the rain moves in, the air pressure begins to fall. Cats are able to detect slight changes in air pressure as a result of their very sensitive inner ears, which also allow them to land upright when falling. The cat will rub its ears for relief. When

Grandma saw a cat rubbing up against the table legs she knew that there was going to be a change in the weather, and she was right. Kitty was doing that to relieve some of the pain in her ear canal brought on by changing barometric pressure. I've also heard people say that when a cat is restless and darts from place to place, you can expect high winds. Low atmospheric pressure, a common precursor of stormy weather, often makes cats nervous and restless. So the next time Fluffy is freaking out, you might think twice about heading out.

A rainbow in the morning is the shepherd's warning; a rainbow at night is the shepherd's delight.

. . . .

D EPENDING ON HOW old you are, it might be Roy G. Biv or "right on your great big veranda." Those are two fun ways that kids around the world are taught to remember the order of colours in the rainbow: red, orange, yellow, green, blue, indigo, and violet.

Everything stops when someone spots a magnificent arc of coloured light in the sky. The first to really try to describe the rainbow was Aristotle. He thought that the rainbow was caused by the reflection of sunlight in the clouds. Today, we know the specifics. Sunlight is reflected by raindrops. We see the sun's light as "white light." But in fact, it isn't white at all; it's many different colour wavelengths. The main colours it gives off are red, orange, yellow, green, blue, indigo, and violet—the colours of the spectrum.

While the science behind the rainbow might be complicated, Grandma believed that its message was fairly simple: "A rainbow in the morning is the shepherd's warning; a rainbow at night is the shepherd's delight."

When you see sunshine while it's raining, turn your back to the sun and that's where you'll find the rainbow, and hopefully a pot of gold. If it's morning, the sun will be in the east. With your back to the sun, you'll be facing the rain shower in the western sky. Since most of our weather travels from west to east, that rain is moving in—the shepherd's warning.

Conversely, if the sun shower is happening late in the day, the sun will be in the west. With your back to the sun, you'll be facing the shower in the eastern sky. That shower will be on its way out—shepherd's delight.

By the way, in case you plan on chasing that elusive pot of gold, you can never actually reach the end of a rainbow. As you move, the rainbow that your eyes see moves as well, because the raindrops are at different spots in the atmosphere. The rainbow, then, will always "move away" at the same rate that you are moving.

A cow with its tail to the east makes the weather least; a cow with its tail to the west makes the weather best.

T HERE'S NO DISPUTING the fact that farm life is not easy, but if you're a kid there's nothing like it. Most of the time we thought it was the coolest place on earth, but if we lost sight of that, we were quickly reminded

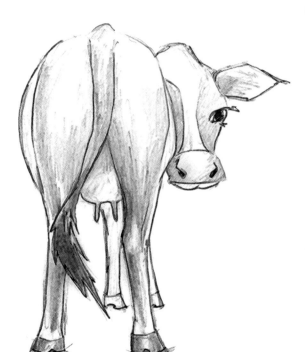

of just how neat it was when our cousins visited from the city. They loved to go running through the fields; their backyards were so tiny in comparison. They climbed the ladder into the hay mow and played hide-and-seek for hours. They couldn't wait to cuddle in the straw with the kittens or chase the hens around. But what about the cows? The cows were our bread and butter. Didn't they deserve a little attention too?

Well, our cows were certainly not neglected. I can tell you that Mom and Dad loved their cows, and Grandma loved to watch them. Sure they're big and a little clumsy looking, but they are sensitive creatures—at least sensitive to the weather. After careful observation, our ancestors concluded that "A cow with its tail to the east makes the weather least; a cow with its tail to the west makes the weather best."

In nature you'll find that most animals graze with their tail to the wind. It's a natural instinct. This way the animal can face and see an invader coming, but if the threat is approaching from behind, the wind will carry its scent to the animal's nose. This way, all the bases are covered. As a general rule, fair and dry weather occurs when the wind is blowing from the west. Wet and inclement weather occurs when the wind is blowing from the east. So, if the cows stand with their tails to the west, the weather is "best." If they're facing away from the wind when it is from the east, then you're more likely to have stormy or wet weather. So, in as much as an east wind is a rain wind and a west wind is a fair wind, the grazing animal's tail becomes a weather sign.

The "tock, tock, tock" of a woodpecker announces rain.

SOMETIMES WITHOUT EVEN realizing it, we go through life taking advantage of the most amazing things around us. I discovered that at a pretty young age. One summer, I had the opportunity to visit my godmother in the city. I was going to spend a few nights away from the farm for the very first time. I was so excited about all the new things I encountered on my way to Tante Jeannette's house that I was exhausted by the end of the day. But that was okay, because I had my very own room. That first night I slept like a log. I woke up at sunrise—must have been the farm girl in me. Nobody else was up yet so I just stayed in bed, listening. I heard a few cars go by, but otherwise, nothing. Where were all the birds? Every morning on the farm, I would wake up to the most incredible symphony of song outside my bedroom window. Some mornings, Mom would come in and point out what kind of birds were out there. While the songbirds were my favourite, there were other intriguing birds around. Every once in a while I'd wake up to the distinctive "tock, tock, tock" of a woodpecker.

In many European cultures, the woodpecker is considered a weather prophet; its drumming indicates forthcoming changes. In our home, the woodpecker announced rain. If Grandma had laundry on the clothesline, in it came; she believed that rain wasn't far off if the woodpecker was drumming. Another good observation by Grandma. It's believed that insects can "sense" the onset of wet weather and make plans before us humans do. In monsoon and rainy areas, it's often observed that prior to an onslaught of rain, some buildings are invaded by insects looking for shelter. Insects also take cover by tucking themselves into the crevices of tree bark. I guess you could say that it's the insects and not the woodpecker doing the forecasting here. Regardless, while the woodpecker was feverishly tapping away at the tree, feasting on the insects, Grandma was counting down to the start of the rain. Oh, and if you ever find a woodpecker feather on the ground, pick it up and take it inside. Woodpecker feathers are believed to guard against lightning.

The louder the train whistle the closer the rain.

I F SOMEONE ASKED me to rhyme off a list of sounds that I associate with growing up on the farm it would look something like this: the "cock-a-doodle-doo" of a rooster, the echoing bark of a dog a concession away, the melancholy "moo" of a dairy cow, and a train whistle. That last one might seem a little out of place, but the CN train line ran about a half-kilometre south of our property. The tracks ran behind our sugar bush so we couldn't see the train, but we could certainly hear it—some days better than others. Grandma was always quick to remark when the sound of the whistle seemed especially loud. She knew it wouldn't be long before the rain moved in. She couldn't tell you why, but she didn't leave clothes on the line when it sounded like the train was behind the barn.

That was a good call on her part. If far-away sounds, like a train whistle or cars on a distant highway, appear louder than usual, then rain or snow is coming. The higher the humidity, the better sound travels. Bone-dry air greatly attenuates sound; conversely, humidity is one of several atmospheric factors affecting the propagation of sound. Humid air is less dense because water molecules have less mass than average air molecules do. So, humidity in the air makes the air lighter. However, humidity is not as prominent a factor in affecting low-frequency sound as the bending of sound waves by wind and the temperature profile of the air. During fair weather, sound waves travel upward into the atmosphere. As they rise, they dissipate fairly quickly and have a shorter range. On a cloudy, humid day, sound waves bend back toward the earth and travel further. Grandma once told me that some people she knew gauged the chances of rain by the clarity with which they heard church bells. Whether you're listening for bells or whistles, remember: the farther the sound travels, the nearer the rain.

If you step on a spider, you'll make it rain.

D O YOU REMEMBER your very first record? I know, I'm dating myself, but I still think vinyl was the best. One spring, the Easter Bunny brought my sister and me a K-tel album called *Get it On*. The bunny hid the record in the clothes drier and we found it before Easter, but we never told anyone. There were a few great songs on that record, but my favourite was "Spiders and Snakes" by Jim Stafford. I loved the song but didn't understand why anyone would not like spiders. I realize now that arachnophobia is quite real, but back then I didn't know anyone who was afraid of spiders. In fact, Grandma admired them. She was always very careful not to step on one, especially during haying season. She believed that stepping on a spider would make it rain.

Now this is an interesting one. I've found no direct link between a squashed spider and a forecast of rain, but I can see how this observation turned into a popular weather saying. Spiders become more active just before it rains. They sense the rise in the humidity in the air and scurry around to take shelter before the rain moves in. In many cases, these spiders move into your home just before the summer rains. So, with all these spiders rushing around before the rain starts, the odds of seeing one and, yes, stepping on one, increase.

When a rooster goes crowing to bed, he will rise with a watery head.

COWS AND CROPS paid the bills on the farm, but over the years we dabbled with other interests. For a little while we raised a few pigs. For many years we had laying hens, and of course a few colourful roosters. My favourite was the Rhode Island Red. What a showboat he was. Aside from looking great, the rooster was a predictable morning alarm. Roosters, like many other birds, sing at sunrise—but it is the males of the species who sing the loudest and most frequently. No one knows all of the reasons why birds sing, but some of their activity can be explained. We do know that birds sing to attract mates. They also use their song to mark territory from intruders, and to warn other members of the flock against these intruders. It is also used to re-establish communications with the flock after night has passed. And like people, not all birds wake up sounding great. Some species don't sing very well when they first wake up, but improve after a few hours' practice.

So back to the rooster. According to Grandma, if the rooster crowed before going to bed, there was rain on the way. While that's a very popular weather expression, scientists have never been able to say why. I know that we had a few rooters who crowed in the morning, but we also had a few who crowed whenever they liked, day or night. According to experts, the idea that roosters crow only at dawn is a misconception. The reason roosters rarely crow at night is because they are diurnal animals that sleep at night. If a rooster does crow at night, any number of factors could be to blame. He could be sick, he may sense a predator, or he may just be feeling a bit antsy. Studies have proven that many animals can sense bad weather, and it is possible that some roosters do crow if rain is on the way, but the correlation certainly isn't as cut and dried as this myth suggests. So it's safe to say that if the rooster goes crowing to bed, he might be sensing a threat, but it's not likely a threat of rain.

If woolly fleeces spread the heavenly way, be sure no rain disturbs the summer day.

I F SOMEONE ACCUSES you of walking around with your head in the clouds, don't be offended. Grandma used to tell me that you can learn a lot more from looking up than looking down. Not only are the cloud shapes interesting, but they can give you a very good idea of what's cooking up there. Being able to predict the weather by observing cloud formation is a skill that is somewhat lost on us modern humans. Most of us can easily look at a cloud and see angels or ice cream cones, but very few of us can look at clouds and see the approaching cold front.

Fortunately, being able to predict the weather with the clouds is easier than you might think. Our ancestors coined a few cute weather rhymes that can walk you through it.

One of my favourites is: "If woolly fleeces spread the heavenly way, be sure no rain disturbs the summer day." Here, "woolly fleeces" refers to a flock of sleeping sheep. The cumulus clouds do look an awful lot like sheep or maybe cotton batten, but sheep are more fun. Afternoon cumulus clouds signal the presence of a high pressure system and fair weather with a very slight chance of rain.

The second part of the weather saying goes like this: "If cumulus clouds are smaller at sunset than at noon, expect fair weather." This too is quite accurate. The smaller, late-day clouds indicate that their formation was due to daytime solar heating and not an advancing low pressure system. Having said that, there are limits to the growth of these cumuli before the forecast changes: "When clouds appear like rocks and towers, the earth's refreshed with frequent showers." If the small puffs of cotton batten grow into large rocks and tall towers due to solar heating and extra moisture in the air, localized showers may very well be in the offing.

And finally, here's Grandma's favourite: "Mountains in the morning, fountains in the evening." Here, "mountains" refers to tall, billowing cumulus clouds. When you look up and see these overhead in the morning, atmospheric conditions are ripe for the development of cumulonimbus clouds. The cumulonimbus cloud, also known as a CB cloud, will often produce late-afternoon or evening thunderstorms. Armed with this valuable information, you will no doubt wow your family and friends at the next backyard BBQ.

When chickens roost during the day, wet weather is on the way.

WHEN PEOPLE ASK me what it was like to grow up on a farm, I quickly answer, "It was an education." It really was. Every time I turned around, there was a lesson to be learned. On the land we learned about soil erosion, crop rotation, and the importance of rain, yes, even on weekends. In the main barn, we learned about lactation, feeding plans, and animal care. Next to the barn, a curious little building housed the chickens. We always kept just enough hens and roosters to supply our eggs.

Chickens are odd creatures. They're not cuddly or playful like cats or dogs. At first glance they don't seem to have much personality, at least that was my assessment, but Mom would tell you otherwise. Mom loved her hens. Her favourites were the little bantam, or as we called them, "banty" hens. They were quite fond of Mom. I remember seeing a couple of them follow her around when she went in to feed them. On our farm the hens' main purpose was to lay eggs.

Most of us eat eggs on a fairly regular basis. They're high in protein, quick and easy to prepare, and very tasty. But did you ever wonder about the process? Now, I don't mean in a large poultry operation, but on a smaller scale, like on the family farm. It's pretty involved. The chickens are kept in a chicken coop. They need a roost; this is a place for the chickens to sleep. By instinct, chickens want to roost or go to bed in the highest point available and be gathered in a group for protection and warmth while they sleep. Once it starts to get dark, one by one the chickens will go into the coop, get up on the roost, and settle down for the night.

Sometimes the chickens would roost during the day. According to Grandma, that didn't mean they were under the weather; instead, they were reacting to the weather. Grandma said, "When chickens roost during the day, wet weather is on the way." This little weather gem has some degree of truth behind it. It would appear that the chickens are fooled by the thickening and darkening clouds that move in ahead of the rain. The dimming daylight tricks them into believing that night is approaching, and for that reason they go to roost. The next time you sit down to a delicious plate of scrambled eggs, I hope you share this little story with someone at the breakfast table.

Red sky at night, sailor's delight;
red sky at morning, sailors take warning.

B ACK ON THE farm, the days started very early in the morning. For years
I held on to the belief that "early to bed and early to rise makes a man
health, wealthy, and wise." I'm not sure now, but I am convinced that those
who sleep in miss out on something truly magnificent: daybreak. Sunsets can
be breathtakingly beautiful, but there's something almost religious about a
bloodstained sky as the morning light brightens to welcome a new day. Of
course not all sunrises are red, and that's a good thing. This is nicely expressed
in the most popular weather saying in the world—and accurate too. "Red sky
at night, sailor's delight; red sky at morning, sailors take warning" is right on
ninety-seven per cent of the time. The reason is not nearly as magical as the
display, but it's based in science and is interesting nonetheless.

In order to understand why this old weather adage can so
accurately predict the weather, we should start by looking at what causes the
colours in the sky. The colours we see are a result of sunlight, which is white
light, being split into the colours of the spectrum as it passes through the
atmosphere and ricochets off the water vapour and particles. During sunrise
and sunset the sun is low in the sky, and it transmits light through the thickest
part of the atmosphere. A red sky suggests an atmosphere loaded with dust
and moisture particles. We see the red because red wavelengths, the longest
in the colour spectrum, are breaking through the atmosphere; the shorter
wavelengths, such as blue, are scattered and broken up.

Okay, now we have to remember that our weather usually moves from west to east, steered by the westerly trade winds, therefore most of our storm systems move in from the west. When we see a red sky at night, this means that the sun, setting in the western sky, is sending its light through a high concentration of dust particles. This usually indicates an area of high pressure, stable air, and fair weather. If the high pressure system is to our west, chances are it will be overhead by morning—sailor's delight. A red sunrise reflects the dust particles of a fair weather system that has already moved through; it's east of us, on its way out. For every high there's a low, so we can likely expect some unsettled weather before long—sailors take warning.

And there it is, the most talked about weather expression on the planet. It doesn't disappoint very often but when it does, it usually happens at the east coast. There, every once in a while, systems retrograde or back up. An east-to-west weather pattern makes a mess of the most accurate saying of them all.

If there's enough blue sky to mend a pair of Dutchman's breeches, the weather will be fine.

WHEN GRANDMA WENT for a walk, she didn't spend half her time texting or fiddling with her iPod—she was looking around. I loved to go for walks with Grandma. I remember one such walk. We were on our way to pick apples along the line fence. The morning had been quite dark and cloudy so we wondered if we would get wet before returning to the house with

CINDY DAY

our harvest. Grandma gazed up at the sky and said, "Look, there's enough blue sky to mend a Dutchman's breeches. We'll be fine."

I don't know how Grandma got a hold of this one. The expression is part of traditional seagoing weather lore. It was believed that in bad weather, two patches of blue sky was a hopeful sign as long as the patches were big enough to "mend a pair of Dutchman's breeches." Many years ago, sailors wore wide trousers, and Dutch sailors were known to wear even wider trousers and yes, you guessed it, their trousers were blue, like the sky on a clear day. Somehow that maritime weather expression made its way to the farm. Whenever Grandma saw two patches of blue that appeared in the middle of a stormy sky, she believed that the storm was breaking up.

There's definitely some truth to this one. The stratocumulus clouds are large masses or rolls of dark cloud that frequently cover the whole sky. They can give the sky a wavy appearance. It is not a very thick layer of cloud, and occasionally blue sky is visible between the rolls or waves. The breaks in the cloud cover indicate the presence of drier air above and usually indicate a clearing trend. When those patches of blue are large enough to mend a pair of Dutchman's breeches, the rest of the day will surely be pleasant.

Tipped moon, wet; cupped moon, dry.

W**HAT AM I?** I turn while you turn; the rotations on my axis keep exact time with my revolutions around your globe; I accompany you as you circle the sun, always facing you, never turning my back on you. Some believe I can cast spells and bring good luck; I'm known to influence the tides and induce labour, but I do not control your weather.

I am the moon.

That last clue might have thrown you off a little. For centuries, humans have been mystified by the moon—observing its light, studying its phases—but over time there has never been a conclusive link between the moon and the weather. Don't get me wrong: the moon has a very real impact on us. Many gardeners still plant certain crops by the phases of the moon. Grandma would tell you to plant crops that are valuable because of the parts that grow above ground—such as corn and wheat—while the moon is waxing or growing larger. Neap and spring tides, the highest and lowest tides of the seasons, are directly influenced by the phases of the moon.

Having said that, it doesn't mean that the folklore isn't out there. Grandma was always quick to point out that if the big dipper was pointing downward, water could pour out and it would rain. If it was tipped upward, we were entering a dry spell. This one is a mystery to me. I can't find any correlation between the position of the stars and the weather. Stars change their position over the seasons. As the Earth orbits the sun, and as the seasons change from spring to summer to fall and to winter, the area of the sky you can see at night sweeps around the sun in a circle. That would give us a changing perspective on the position of the stars in the sky.

Here's another weather expression from the night sky. It's similar to the previous one and has become quite popular over the years: "Tipped moon, wet; cupped moon, dry." The way a crescent moon's horns point really has nothing to do with the weather. This is purely a function of the moon's orbit. This little poem brings the point home quite nicely: "The moon and the weather may change together, but a change in the moon can't change the weather."

If you boil the potatoes dry, it's going to rain.

BACK ON THE farm, everybody spent a lot of time in the kitchen. It wasn't unusual to see Mom mending a pair of socks at the table in the evening, or dad reading the newspaper after the mail went by. When I was young, I even did my homework on the old melamine table. Mom would get supper started then head out to the barn to do some chores with Dad. Grandma stayed in to keep an eye on us kids and watch the pots on the stove. Sometimes, Grandma would sit with us and we'd lose track of time. Before you knew it, that awful smell of scorched potatoes would fill the air. The next thing we knew she was going on about how there was rain coming.

When I started to study meteorology that was one of the first things I wanted to look into. This one's for all the cooks out there. The boiling point of a liquid varies depending on the surrounding air pressure. A liquid in a high pressure environment has a higher boiling point. When the pressure is low, as is the case ahead of an approaching storm, the water will boil a little more quickly. Think about it this way: when water heats up, the molecules move faster and move farther apart, eventually becoming steam. When you increase the pressure on the water you are, in a way, holding the molecules together more tightly. This helps keep them from moving apart so, it requires more heat to get them to "break free." So, the higher the air pressure, the higher the boiling point, the longer it will take to get the water boiling. Conversely, the lower the air pressure, the lower the boiling point and the more quickly the pot of water will boil dry. Since Grandma always boiled her potatoes in the same pot, she knew exactly how long it took, unless the outside pressure was falling. That's when Grandma scorched the potatoes, and that's how she knew there was rain on the way.

Ring around the sun or moon, rain or snow upon us soon.

G RANDMA LOVED TO read, but
life on the farm was very busy.
She didn't spend a lot of time sitting with a
book, but she always made time to read the sky.
She use to say that Mother Nature sends us signs of
things to come; we just have to know what to look for. She
was right.

One of those signs was the infamous ring around the sun or the
moon. That ring is called a "halo." During the day, it's a result of the sun's rays
reflecting off the ice crystals that make up cirrus-type clouds. The conditions
have to be just right. The ice crystals must be hexagonal and lying horizontally.
This perfect scenario doesn't happen that often. When the ice crystals are in
that ideal position, the sun's rays reflect off them in such a way as to create a ring
around the sun. The same thing happens with the moonlight reflecting off ice
crystals. Most halos appear as bright white rings, but in some cases, the scattering
of light as it passes through ice crystals can cause a halo to have some colour.

So how is this a sign of rain or snow? The icy upper-level cirrus cloud
needed for the halo to form is one of the first clouds to lead an advancing
weather maker; that cirrus cloud appears anywhere from twenty-four to thirty-
six hours ahead of the rain or snow.

When Grandma noticed a halo in the sky she would say: "Ring around the
sun or moon, rain or snow upon us soon."

When seagulls gather over land, a change of weather is close at hand.

· ·

W̶E'VE ALL HEARD the expression "a bird's eye view." Imagine how cool it would be to fly high toward an incoming storm. You'd be able to see the cloud layers lowering as the system approached and you'd get an early sampling of the inbound winds too. Oh, to be a seagull. I guess you could say seagulls are our eyes ahead of an offshore storm. You might not know this, but seagulls are not especially fond of standing or walking. They're naturally at home in flight, and where they can, they'll even sleep on the water rather than have to stand on land. However, seagulls, like people, find gusty, turbulent

wind difficult to contend with. Such conditions, typical ahead of a low pressure system, also make the water quite choppy. So when the winds start to gust and the water begins to churn, these seabirds head for land. After decades of careful observation, this little gem was born: "When seagulls gather over land, a change of weather is close at hand." Can you blame them? To be fair though, seagulls huddled on the ground are really not a predictor of bad weather as much as they are a sign that the weather is already bad, at least offshore.

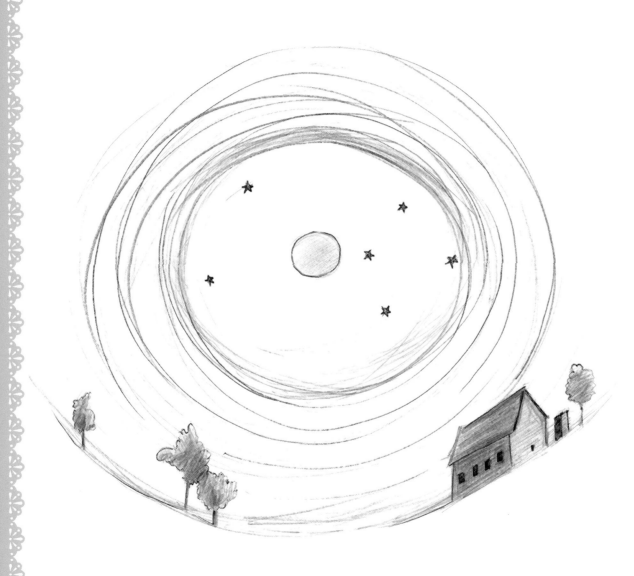

CINDY DAY

Stars inside the halo.

· · · · · · · · · · · · · · · · · · · ·

ONCE YOU'VE SPOTTED a lunar halo in the night sky, your forecasting fun doesn't end there. If you look closely, you might see a few stars inside the ring. Grandma believed that the number of stars inside the circle was the same as the number of days before the rain or snow would move in.

Why? Well, you have a much better chance of seeing lots of stars inside the ring if the sky is very clear. In order for the ring to form in the first place, there must be a thin veil of cirro-stratus clouds. Those clouds are made up of wispy sheets of ice crystals, as high as five kilometres above the ground. They're also forerunner clouds; they reach out ahead of an advancing weather system. The density of those ice crystals increases as the weather system approaches. That increased density will make the sky look quite hazy or milky, and block out some of the starlight. So the more stars you see, the more time you have before the rain arrives. Fewer stars indicate that the system is not far off. So count the stars inside the ring and wait. Grandma wasn't wrong very often!

When windows won't open and the salt clogs the shaker, the weather will favour the umbrella maker.

YOU MIGHT HAVE noticed that a lot of our weather lore pertains to rain. In the early and mid-1900s, people spent so much more time outdoors. Even if work was inside, getting there could be a challenge if the weather was inclement. People travelled by horse and buggy, roads were not paved so mud was often an issue, and kids didn't have the luxury of school buses. It was important to know if rain was on the way.

Here is one of my favourite rainy-day weather sayings: "When windows won't open and the salt clogs the shaker, the weather will favour the umbrella maker."

Back on the farm, we had old wooden windows that were difficult enough to open at the best of times, but before it rained, Grandma always commented on how tight the windows were. Moisture in the air causes wood to swell, making doors and windows sticky. That moisture or humidity in the air rises ahead of the rain, so even before the rain had arrived, Grandma knew it was coming.

As far as the salt goes, she took matters into her own hands. After the cold of winter had passed, Grandma would add rice to the salt shakers. She knew that salt was a very effective absorber of moisture. She also knew that the humidity would be higher during the summer months and before long the salt would start to clump together. That's where the rice comes in; after a few good shakes, the rice would help break up the lumps in the shaker.